VOLUME SEVENTY NINE

ADVANCES IN BOTANICAL RESEARCH

Artificial Photosynthesis

ADVANCES IN BOTANICAL RESEARCH

Series Editors

VOLUME SEVENTY NINE

Advances in BOTANICAL RESEARCH

Artificial Photosynthesis

Volume Editor

ROBERT BRUNO

Institut de Biologie et de Technologies de Saclay, CEA, UMR 8221 CNRS, Université Paris Sud, CEA Saclay, Paris, France

ELSEVIER

AMSTERDAM • BOSTON • HEIDELBERG • LONDON
NEW YORK • OXFORD • PARIS • SAN DIEGO
SAN FRANCISCO • SINGAPORE • SYDNEY • TOKYO

Academic Press is an imprint of Elsevier

Academic Press is an imprint of Elsevier
125 London Wall, London EC2Y 5AS, UK
The Boulevard, Langford Lane, Kidlington, Oxford OX5 1GB, UK
50 Hampshire Street, 5th Floor, Cambridge, MA 02139, USA
525 B Street, Suite 1800, San Diego, CA 92101-4495, USA

First edition 2016

ISBN: 978-0-12-803289-3
ISSN: 0065-2296

For information on all Academic Press publications
visit our website at https://www.elsevier.com

Publisher: Zoe Kruze
Acquisition Editor: Kirsten Shankland
Editorial Project Manager: Hannah Colford
Production Project Manager: Magesh Kumar Mahalingam
Designer: Matthew Limbert

Typeset by TNQ Books and Journals

CONTENTS

CONTRIBUTORS

D. Abramavicius
Vilnius University, Vilnius, Lithuania

V. Artero
University of Grenoble Alpes, Grenoble, France; CNRS, LCBM and SPrAM, Grenoble, France; CEA, BIG/CBM and INAC-SPrAM, Grenoble, France

A. Aukauloo
Institut de Chimie Moléculaire et des Matériaux d'Orsay, UMR-CNRS 8182, Université Paris-Sud XI, Orsay, France; Service de Bioénergétique, Biologie Structurale et Mécanismes (SB_2SM), CEA, iBiTec-S; Biochimie Biophysique et Biologie Structurale (B_3S), I_2BC, Gif-sur-Yvette, France

F. Chandezon
University of Grenoble Alpes, Grenoble, France; CNRS, LCBM and SPrAM, Grenoble, France; CEA, BIG/CBM and INAC-SPrAM, Grenoble, France

Q. Chen
University of Amsterdam, Amsterdam, The Netherlands

D.T. Co
Northwestern University, Evanston, IL, United States

B. Dietzek
Leibniz Institute of Photonic Technology, Jena, Germany

S. Fiechter
Helmholtz-Zentrum Berlin für Materialien und Energie GmbH, Berlin, Germany

R. van Grondelle
Vrije Universiteit, Amsterdam, The Netherlands

D. Gust
Arizona State University, Tempe, AZ, United States

K.J. Hellingwerf
University of Amsterdam, Amsterdam, The Netherlands

J.T.M. Kennis
Vrije Universiteit, Amsterdam, The Netherlands

W. Leibl
Service de Bioénergétique, Biologie Structurale et Mécanismes (SB_2SM), CEA, iBiTec-S; Biochimie Biophysique et Biologie Structurale (B_3S), I_2BC, Gif-sur-Yvette, France

D. Montesarchio
University of Amsterdam, Amsterdam, The Netherlands

A.L. Moore
Arizona State University, Tempe, AZ, United States

T.A. Moore
Arizona State University, Tempe, AZ, United States

S. Pillai
Arizona State University, Tempe, AZ, United States

J. Ravensbergen
Vrije Universiteit, Amsterdam, The Netherlands

J.J. Snellenburg
Vrije Universiteit, Amsterdam, The Netherlands

I.H.M. van Stokkum
Vrije Universiteit, Amsterdam, The Netherlands

L. Valkunas
Vilnius University, Vilnius, Lithuania; Center for Physical Sciences and Technology, Vilnius, Lithuania

FOREWORD

What is the aim of this book ?

For more than two centuries, we have had access to cheap, abundant, energy through coal, then through petroleum, and for long the only concern about the use of fossil energies was the pollution they induced and the fact that they would come to an end. In 'The Limits of Growth', report from the Club of Rome written in 1972, where the future challenges of mankind for the next centuries were addressed in detail, although the exponential increase of CO_2 concentration in the Earth's atmosphere was discussed, the potential effects on the planet climate of the human activity were just very briefly mentioned. The effect of human activity on Earth's global climate became a public issue only in the late 1980s, leading to the creation of the Intergovernmental Panel on Climate Change (IPCC). It became clearer and clearer since that one of the major challenges humanity has to face at short term is the progressive global warming of the planet induced by the increase of CO_2 in the atmosphere and its potential devastating consequences. Curiously, the relationship between CO_2 concentration and the temperatures at the surface of the planet has been addressed as soon as at the end of the 19th century by Arrhenius, in a quite detailed way, but it took nearly a century for the drift of mean temperature to become clear enough to be widely recognized. If nowadays, still a few voices contest the global warming, every year the measured planet temperatures confirm it as an evidence.

We presently face the absolute constraint of limiting the amount of CO_2 we emit in the atmosphere, and still a sizeable fraction of the human population has no or limited access to energy. The most optimistic estimations predict mankind to need about 30 TW per year in 2050, ie, twice more as we use now. In the present state of technology, the only way to achieve such an energy production without giving rise to a major climatic catastrophe is to learn how to use solar energy. The solar energy reaching the surface of the Earth is virtually unlimited as compared to human needs for energy. Thanks to photovoltaics we are able to transform it into electricity with increasing yields; however, we are not able yet to store this electricity into a stable form with high efficiency. Plants and more generally photosynthetic organisms have transformed directly solar energy into biomass for billion years, and they store every year about twice what mankind will need in 2050. Among the most promising approaches to solve the forthcoming energy crisis, one is to develop devices which would, as photosynthetic organisms do, directly transform the solar energy into fuels. This is what is called 'artificial photosynthesis'.

Artificial photosynthesis can be achieved in a number of ways, from improving natural photosynthesis to solid-state approaches, covers a wide range of scientific fields involves theoretical as well as experimental work. The aim of this book is to summarize the different aspects of the artificial photosynthesis field, which, I strongly believe, will become one of the essential keys for the future of energy.

CHAPTER ONE

An Illustrative History of Artificial Photosynthesis

D. Gust
Arizona State University, Tempe, AZ, United States
E-mail: gust@asu.edu

Contents

Abstract

This chapter presents a brief history of molecule-based artificial photosynthesis research. The emphasis is on early work on the subject beginning in the late 1970s, but the evolution of the field up through the present day is discussed. The main subjects covered are artificial reaction centres and photoinduced electron transfer, artificial photosynthetic antennas, mimicry of photosynthetic photoprotection and

Advances in Botanical Research, Volume 79
ISSN 0065-2296
http://dx.doi.org/10.1016/bs.abr.2016.02.004

photoregulation and the construction of systems that couple these processes to the production of fuels.

1. INTRODUCTION

This chapter is an overview of the development of the science of artificial photosynthesis with an emphasis on the beginnings of the field. This is not a review article: the presentation will be brief and necessarily illustrative, rather than comprehensive. Many of the examples presented here that typify the various stages of development of the field are drawn from my research with my colleagues Professors Thomas A. Moore and Ana L. Moore at Arizona State University, but the work of other investigators would serve just as well in many cases. Keep in mind the following comment from the historian Eugen Weber: 'If cats could write history, their history would be mostly about cats' (Weber, 1989).

1.1 What is Artificial Photosynthesis?

Artificial photosynthesis has been characterized in many different ways, with the definition depending on the experience, predilections and motives of the definer. Because natural photosynthesis is basically a process for converting sunlight into other, more useful forms of energy, artificial photosynthesis is sometimes taken to be anything that carries out a similar function, whether the product be electricity or a fuel, whether the materials be extended solids or molecular, and whether these materials include organic or inorganic molecules or metal—organic complexes. For the purposes of this brief history, we will define artificial photosynthesis relatively narrowly.

Natural photosynthesis converts sunlight into electrochemical potential energy and uses this energy to create fuels such as carbohydrate, lipids or even hydrogen gas. The components of the system are (bio)organic molecules and metal—organic complexes, and the process is photochemical. Here, we define artificial photosynthetic systems similarly: photochemical systems that use at least some organic or metal—organic molecular materials to harvest sunlight and convert it to electrochemical potential, with an ultimate goal of fuel production. This definition restricts our discussion to a manageable subset of results.

1.2 Why is Artificial Photosynthesis Important?

Later in this Chapter, I will show through examples why artificial photosynthesis research is interesting. Here, I will try to convince the reader that it is also important – for science and for society. From the scientific point of view, artificial photosynthesis research has helped spur the development of many important aspects of modern photochemistry. These include interchromophore energy and electron transfer, which as we shall see are integral to mimicry of the fundamental processes of natural photosynthesis. The design and synthesis of many new chromophores, electron donors and electron acceptors have occurred in the context of artificial photosynthesis, as has the development of new ultrafast spectroscopic tools, X-ray crystallographic methods of structure determination and theoretical methods. Artificial photosynthesis research not only draws on the natural analogue for inspiration, but also helps elucidate the details of natural photosynthesis because artificial photosynthetic model compounds abstract important aspects of the natural process into simpler, better defined, more easily studied systems wherein parameters may be more easily quantified and varied.

From a societal point of view, artificial photosynthesis holds promise for lessening our dependence on fossil fuels for technological energy. To a large extent, modern society is based on energy from fossil fuels. Fossil fuels power everything from coal stoves to cell phones, and have made the industrial revolution and ensuing development possible. However, there are major problems arising from fossil fuel use. The ultimate, unavoidable problem is that they will eventually be used up. This likely would not happen for several hundred years, but several hundred years is a short time in terms of the history of humans, and we hope in terms of the future of society. A replacement must be found. Second, using fossil fuels generates carbon dioxide, which is a greenhouse gas demonstrated to be important in climate change. The extraction of fossil fuels leads to other forms of pollution as well, including oil spills, coal mine fires, methane release and environmental degradation due to mining. Finally, fossil fuels are not generally distributed around the earth in the same way that people are distributed. Competition for fuel deposits leads to serious economic and political problems.

There is ample sunlight reaching the surface of the earth to meet all human requirements for the imaginable future. Photosynthesis is a natural technology operating at huge scale which has been converting sunlight into energy stored in fuels (including the fossil fuels) for several billion years.

The attraction of developing artificial photosynthesis to make fuels that can replace the natural fossil fuels is obvious.

2. EARLY MUSINGS AND PROMISING FINDINGS

Humans have recognized the importance of the sun as a heat and energy source for eons, and the sun plays a major role in many cultures and religions. Although farmers have undoubtedly understood at some level since antiquity that plants require sunlight, a landmark in the comprehension of natural photosynthesis was the discovery in 1779 by the Dutch scientist Jan Ingenhousz that the green parts of plants release oxygen, and that this happens only under sunlight. By the early 1800s it was understood that plants use sunlight to convert carbon dioxide and water into organic compounds, with oxygen as a by-product. The idea that the sun serves as the energy source for all life, and that photosynthesis is basically conversion of the energy of light into chemical potential energy was enunciated in 1845 by the German physician and physicist Julius Robert von Mayer, who also formulated some early thermodynamic principles. Thus, the broad outlines of photosynthesis were understood by the middle of the nineteenth century, and the term was in use by the beginning of the twentieth century.

The beginnings of an understanding of photosynthesis gave rise to the concept of using the basic science underlying natural photosynthesis as inspiration for human-designed systems for converting solar energy to useful forms — artificial photosynthesis. In 1912, Giacomo Ciamician of the University of Bologna, one of the first modern photochemists, presented a lecture entitled *The Photochemistry of the Future* in New York City. The quote below from the printed version of that lecture (Ciamician, 1912) is likely the earliest and undoubtedly the most lyrical exposition of artificial photosynthesis.

> *On the arid lands there will spring up industrial colonies without smoke and without smokestacks; forests of glass tubes will extend over the plains and glass buildings will rise everywhere; inside of these will take place the photochemical processes that hitherto have been the guarded secret of the plants, but that will have been mastered by human industry which will know how to make them bear even more abundant fruit than nature, for nature is not in a hurry and mankind is. And if in a distant future the supply of coal becomes completely exhausted, civilization will not be checked by that, for life and civilization will continue as long as the sun shines!*

When this was written, neither our understanding of natural photosynthesis nor the state of development of photochemistry and related sciences was anywhere near the levels necessary to design a functional artificial photosynthetic system. In the first half of the twentieth century, much of this groundwork was laid, both in the physical and the biological sciences.

In 1972, Fujishima and Honda reported the construction of a photoelectrochemical cell which split water into molecular oxygen and hydrogen upon illumination of a TiO_2 electrode with light capable of band gap excitation of the semiconductor. The quantum yield was very low, and ultraviolet light, which is almost absent in natural sunlight at the earth's surface, was necessary for band gap excitation, but this experiment, although not artificial photosynthesis by our definition, demonstrated for the first time that direct photochemical splitting of water to produce a fuel was possible.

3. THE BEGINNINGS OF MODERN ARTIFICIAL PHOTOSYNTHESIS: DONOR–ACCEPTOR DYADS

The heart of photosynthesis is photoinduced electron transfer. In this process, light is absorbed by a chromophore and an excited state is produced. The energy from light that is stored in this excited state is used to transfer an electron from a donor moiety (often the chromophore) to an electron acceptor moiety to generate a charge-separated state. Some of the original photon energy is stored in the charge-separated state. By the mid-1970s, photochemists had studied photoinduced electron transfer between separate donor and acceptor molecules in solution using small molecule chromophores that absorbed mostly UV light. These studies helped establish a framework for understanding electron transfer in general, but there were drawbacks. In the first place, electron transfer requires overlap of molecular orbitals, and this in turn requires collisions of separate molecules in solution. Thus, the time scale for electron transfer is limited by diffusion. Because the energetic singlet excited states of organic chromophores are typically very short lived (a few nanoseconds or less), diffusion-limited photoinduced electron transfer is observed mainly with the triplet states of chromophores, which are generally of considerably lower energy content than the excited singlet states and are not used in natural photosynthesis. In addition, the separations and orientations of the donors and acceptors undergoing electron transfer are not well defined, and this typically limits efficiency and makes testing of electron transfer theories difficult.

This limitation can be overcome by covalently linking a chromophore and a suitable electron acceptor (or donor). The linkage defines or at least limits the relative separation of the moieties, and the time constraints of diffusion are no longer an issue. In addition, electronic orbitals of the linkage usually play an important role in the overlap of the orbitals of the donor and acceptor, and thus in determining the rate of electron transfer. By the 1970s, chemists began to synthesize covalently linked donor—acceptor 'dyads' for the study of photoinduced electron transfer.

3.1 Porphyrin—Quinone Dyad Artificial Reaction Centres

In natural photosynthesis, photoinduced electron transfer occurs in a reaction centre. Chlorophyll or bacteriochlorophyll molecules in the reaction centre are excited, and their excited singlet states donate electrons to acceptor moieties to generate energetic charge-separated states. In my view, the first artificial photosynthetic donor—acceptor dyads were reported in 1978 by Kong and Loach at Northwestern University in the United States (Kong & Loach, 1978) and in 1979 by Tabushi and coworkers in Japan. These molecules are shown as **1** and **2** in Fig. 1. Each dyad consists of a porphyrin covalently linked to a quinone. These porphyrins are synthetic molecules, but are closely related to natural chlorins and bacteriochlorins such as the chlorophylls. Like chlorophylls, they are cyclic tetrapyrroles, and have absorption bands scattered throughout the visible spectral region. Thus they can absorb sunlight with reasonable efficiency, although the longest-wavelength absorption (a Q band) typically has a considerably lower extinction coefficient than do comparable bands in chlorophylls. Their first oxidation potentials are not too different from those of chlorophylls, and thus they can act as excited state electron donors in artificial photosynthetic constructs. Unlike chlorophylls, they are readily synthesized and have good stability in solution. Their optical and redox properties may be altered by substitution at various positions on the macrocyclic ring, and by introduction of metals. For these reasons, synthetic porphyrins have now become the most common light absorbers and excited state electron donors in organic artificial photosynthesis research.

The electron acceptor moieties in **1** and **2** are benzoquinone derivatives. Quinones are small, readily prepared, and have tunable redox potentials in the region of interest for artificial photosynthesis. They are also the ultimate electron accepting moieties in natural photosynthetic reaction centres, and have been used as electron acceptors in hundreds of artificial photosynthetic molecules since 1979 (Connolly & Bolton, 1988).

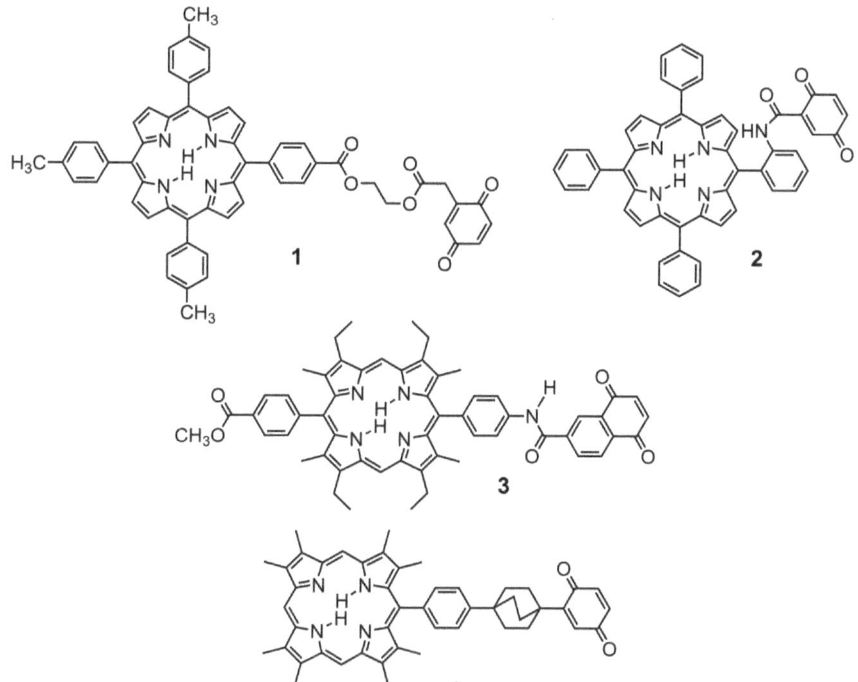

Figure 1 Structures of porphyrin–quinone dyads reported by Kong and Loach (**1**), Tabushi and coworkers (**2**), Kuciauskas et al. (**3**) and Dervan and coworkers (**4**).

Time-resolved spectroscopic experiments reveal the photochemistry of the dyads. Porphyrin (P)–naphthoquinone (NQ) dyad **3**, reported by Kuciauskas et al. (1997) (Fig. 1), will be used to exemplify the light response. Excitation of the porphyrin generates its first excited singlet state ^{1}P-NQ. Based on the absorption and fluorescence emission spectrum of the porphyrin, the energy of this state is 1.97 eV above the ground state. The main pathway for decay of the porphyrin excited state is by photoinduced electron transfer to the quinone to generate a $P^{\bullet +}$-$NQ^{\bullet -}$ charge-separated state. The thermodynamic driving force for this electron transfer is the difference between the energy of the excited state and that of $P^{\bullet +}$-$NQ^{\bullet -}$. The $P^{\bullet +}$-$NQ^{\bullet -}$ state lies about 1.45 eV above the ground state, as estimated from the electrochemically determined redox potentials of the component P and NQ moieties. Thus, the ΔG^0 for electron transfer is −0.52 eV. The rate constant for photoinduced electron transfer can be determined from time-resolved spectroscopy. The first excited singlet state of a model porphyrin for P decays in 9.3 ns by the usual unimolecular processes of internal conversion, fluorescence and intersystem crossing to the triplet state. In **3**, ^{1}P-NQ decays in 102 ps. The dramatic lifetime reduction relative to the model porphyrin is due to photoinduced electron transfer to yield $P^{\bullet+}$-$NQ^{\bullet-}$, and corresponds to a rate constant of $9.7 \times 10^9\ s^{-1}$ and a quantum yield Φ of 0.99.

These results suggest that **3** would be an excellent artificial photosynthetic molecule that converts essentially every photon absorbed into energy stored in a charge-separated state that preserves a substantial fraction of the photon energy as electrochemical potential. However, this is not the whole story. Transient absorption measurements show that the $P^{\bullet+}$-$NQ^{\bullet-}$ state of **3** decays by charge recombination with a time constant of only 1.9 ps. Thus, the charge-separated state returns to the ground state with a rate constant that is greater than its rate constant for formation, and the energy stored in the charge separation is lost as heat. Such a short lifetime would make harvesting the energy of the charge-separated state to do useful chemistry extremely difficult. Although not all porphyrin–quinone dyads show charge recombination that is more rapid than charge separation, they all suffer from very short charge separation lifetimes, typically less than 100 ps, and this makes them both undesirable for practical applications and poor models for natural photosynthesis, where the charge separation in the reaction centre can last for about 1 s.

Although the porphyrin–quinone dyads have limitations as artificial reaction centres, their study by a multitude of researchers has provided a great deal of useful information concerning the effects of structure and environment on electron transfer rates (Gust & Moore, 2000). One illustrative example of many is molecule **4**, which was one of a set of molecules reported by Dervan and coworkers (Joran et al., 1987; Joran, Leland, Geller, Hopfield, & Dervan, 1984; Khundkar, Perry, Hanson, & Dervan, 1994; Leland et al., 1985). These molecules were designed to interpose a relatively rigid linkage between the porphyrin and the quinone, and thus to define better the distance between the donor and acceptor moieties, which is an important factor in controlling the rate of electron transfer. By varying the substituents on the quinone, Dervan and coworkers were able to tune the thermodynamic driving force for photoinduced electron transfer, and transient spectroscopy was used to measure the rate constants for the process.

The results for **4** were interpreted in terms of the theory for electron transfer developed by Marcus (Marcus, 1956; Marcus & Sutin, 1985), Hush (1958, 1961) and Levich (1966), and later evolved by others. This theory and its variants have been a cornerstone for planning and understanding the results of electron transfer experiments in the artificial photosynthetic area. According to Marcus theory, an increase in thermodynamic driving force ($-\Delta G^0$) for electron transfer will result in an increase in rate up to a maximum rate, which is on the time scale of molecular vibrations. This is called the normal region of the Marcus rate versus free energy change curve. At the maximum, the driving force equals the sum of the internal and solvent reorganization energies for electron transfer. When the driving force exceeds the total reorganization energy, further increases result in a decrease in rate constant.

In the Dervan molecules, the rate constants for photoinduced electron transfer were found to increase with driving force in the normal region of the Marcus curve, but molecules with sufficient driving force to reach the inverted region were not accessible. Similar results were reported by Osuka and coworkers in the early 1990s (Asahi et al., 1993; Osuka, Zhang, Maruyama, Yamazaki, & Nishimura, 1992) and by others. The existence of the inverted region predicted by Marcus theory was verified in another class of dyads by Closs and Miller and coworkers (Calcaterra, Closs, & Miller, 1983; Closs & Miller, 1988), following work with unlinked molecules by Miller (Miller, Beitz, & Huddleston, 1984).

3.2 Porphyrin–Fullerene Dyad Artificial Reaction Centres

Although the first artificial photosynthetic dyads used quinones as electron acceptors, other moieties have been investigated since. The initial electron acceptor in photosynthesis is in fact a porphyrin, rather than a quinone, and various porphyrin dyads have been reported in which one porphyrin serves as the electron donor and a second as the acceptor. Substituents tune the redox potentials to favour photoinduced electron transfer. Sessler, Johnson, Creager, Fettinger, and Ibers (1990) have reviewed the early work with porphyrin dyads. As with the porphyrin–quinones, rate constants for photoinduced electron transfer in porphyrin dyads depend on the driving force as predicted by the Marcus theory, although dyads with sufficient driving force to show Marcus inverted behaviour are typically not synthetically accessible (DeGraziano et al., 1994). Aromatic imides and diimides have also been studied extensively as electron acceptors with porphyrins, beginning with reports by Beddard, Cowan and Sanders on dyad **5** (Fig. 2) (Harrison, Pearce, Beddard, Cowan, & Sanders, 1987).

One class of acceptors of exceptional interest is the fullerenes. The first porphyrin–fullerene dyad was reported in 1994 (Liddell et al., 1994), and extremely rapid singlet energy and photoinduced electron transfer were

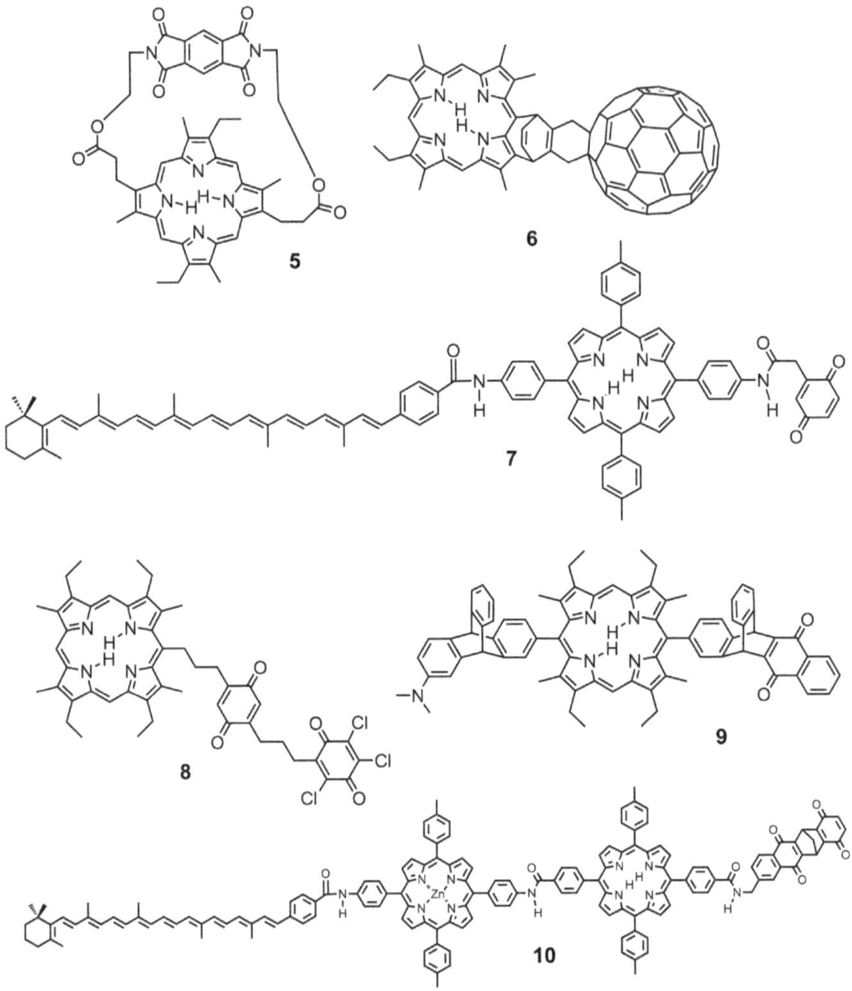

Figure 2 Porphyrin dyads with aromatic imide (**5**) and fullerene (**6**) electron acceptors, three early triad artificial reaction centres (**7, 8, 9**) and molecular pentad **10.**

observed. This report was followed by an avalanche of reports on dyads and more complex molecules with fullerenes as acceptors. Early work was done by Imahori, Sakata and coworkers (Imahori et al., 1995), Boyd, Reed and coworkers (Drovetskaya, Reed, & Boyd, 1995), Schuster and coworkers (Baran, Monaco, Kahn, Schuster, & Wilson, 1997) and Ghiggino, Paddon-Row and coworkers (Bell et al., 1997).

As a set of data on photoinduced electron transfer in porphyrin–fullerene dyads became available, it was observed that by comparison with porphyrin–quinone systems having similar driving force and electron coupling, the porphyrin–fullerene systems tended to demonstrate more rapid photoinduced electron transfer and slower charge recombination. This behaviour has been attributed to smaller internal and solvent reorganization energies for fullerenes (Guldi & Asmus, 1997; Imahori et al., 1996; Liddell et al., 1997), coupled with a smaller sensitivity of the radical anion to solvent stabilization effects (Liddell et al., 1997). Thus artificial photosynthetic constructs with fullerene electron acceptors are more suitable than those with quinones because higher yields of charge-separated states with longer lifetimes can be obtained, and can be obtained under a greater variety of conditions.

4. MORE COMPLEX ARTIFICIAL REACTION CENTRES

Although donor–acceptor dyads of the types discussed above were very important in the early days of artificial photosynthesis research and continue to be useful for some purposes, they all suffer to one extent or another from relatively rapid charge recombination. This failing was a major roadblock in artificial photosynthesis research as pointed out by Melvin Calvin (1978) who was one of the pioneers of artificial photosynthesis research.

Natural photosynthesis, on the other hand, produces very long lived, energetic charge-separated states in high quantum yield. As was evident before X-ray crystal structures of reaction centres were available, and as verified by the first such structure (Deisenhofer, Epp, Miki, Huber, & Michel, 1985; Deisenhofer, Epp, Sinning, & Michel, 1995), reaction centres achieve this result by a multistep electron transfer process. In purple photosynthetic bacteria, for example, excitation of the reaction centre yields the first excited singlet state of a bacteriochlorophyll 'special pair'. Photoinduced electron transfer to a nearby bacteriochlorophyll ensues. The resulting bacteriochlorophyll radical anion in turn transfers an electron to a bacteriopheophytin, which passes charge to a quinone. Finally, the quinone radical anion moves an electron to a second quinone. Each of these electron transfer steps occurs over a relatively short distance, and therefore is fast enough to compete against charge recombination. Yet in the final charge-separated state, the quinone radical anion and special pair radical cation are far apart. The final charge-separation essentially spans the photosynthetic membrane, and this ensures a long lifetime for the state. Thus, photosynthesis prepares an energetic, long-lived charge-separated state in high quantum yield via a series of sequential electron transfers. Each electron transfer step results in some loss of electrochemical potential in order to ensure rapid electron transfer and slow the unproductive reverse electron transfer process.

4.1 Triad Molecules

Inspired by natural photosynthesis, we first implemented this multistep electron transfer approach in molecular triad **7** (Fig. 2), which was reported in 1983 (Gust et al., 1983; Moore et al., 1984). The molecule consists of a tetra-arylporphyrin linked to both a quinone electron acceptor (Q) and a carotenoid polyene (C), which serves as a secondary electron donor. Time-resolved spectroscopic experiments showed that excitation of the porphyrin in benzonitrile solution to yield C-^{1}P-Q was followed by photoinduced electron transfer to the quinone ($k_{ET} = 2 \times 10^9\ s^{-1}$) to yield the C-$P^{\bullet+}$-$Q^{\bullet-}$ charge-separated state with a quantum yield near unity. As with other dyads, this charge-separated state has a large rate constant for charge recombination ($k_{CR} = 6 \times 10^{11}\ s^{-1}$), but competing with charge recombination is a charge shift reaction in which the carotenoid donates an electron to the porphyrin radical cation to produce $C^{\bullet+}$-P-$Q^{\bullet-}$ ($k_{CS} = 1 \times 10^{11}\ s^{-1}$) with a quantum yield of 0.13. In this final state, the radical cation and radical anion are separated by the neutral porphyrin, and this greatly decreases the electronic coupling that leads to charge recombination. Consequently, the final state has a lifetime of 420 ns, which is vastly longer than the few picoseconds lifetime of C-$P^{\bullet+}$-$Q^{\bullet-}$. This triad successfully demonstrated that the multistep electron transfer strategy used by natural photosynthesis may be used in artificial photosynthesis as well.

The year 1983 also saw the report of another triad, molecule **8** in Fig. 2 (Nishitani et al., 1983). This triad was also constructed to mimic the multistep electron transfer processes of photosynthesis. The idea was that an initial photoinduced electron transfer would produce $P^{\bullet+}$-$Q^{\bullet-}$-Q, which would undergo a charge shift to yield $P^{\bullet+}$-Q-$Q^{\bullet-}$. Although the proposed intermediate $P^{\bullet+}$-$Q^{\bullet-}$-Q state was not observed, the final state did indeed form. However, the lifetime of $P^{\bullet+}$-Q-$Q^{\bullet-}$ was only about 300 ps in benzene, which is only moderately longer than the 130 ps lifetime of the $P^{\bullet+}$-$Q^{\bullet-}$ state in a model compound. The relatively short lifetime has been attributed to the flexible polymethylene linkages that join the donor—acceptor moieties (Sakata et al., 1987). Rotation about the bonds in these linkages can serve to bring the porphyrin radical cation and quinone radical anion into close proximity, which facilitates charge recombination.

The successful demonstration of long-lived charge separation with **7** led to the synthesis and study of a large number of triad-type molecules in ensuing years. One of the first, triad **9**, was reported in 1985 (Wasielewski, Niemczyk, Svec, & Pewitt, 1985). This triad consists of a porphyrin primary electron donor, a quinone acceptor and an aniline secondary donor. The rigid bicyclic linkages between the moieties prevent folding motions of the type postulated for **8**. The photochemical pathways for **9** are similar to those for **7**, and the final state had a lifetime in butyronitrile of 2.45 μs.

4.2 More Complex Assemblies

4.2.1 Quinone-Based Artificial Reaction Centres

If two electron transfer steps are good, are more better? In the 1980s and early 1990s, this question was investigated in many laboratories, especially in the United States and Japan. Molecular pentad **10** illustrates what can be achieved (Gust et al., 1990; Gust, Moore, Moore, Machperson, et al., 1993). The pentad consists of a free base porphyrin P bearing a zinc porphyrin Pz and a diquinone moiety (Q—Q). The zinc porphyrin is linked to a carotenoid. The complex electron transfer chemistry of the pentad is shown in Fig. 3. Excitation of a chloroform solution of the pentad at wavelengths where the free base porphyrin absorbs generates C-Pz-^{1}P-Q-Q. This state can also be formed by excitation of the zinc porphyrin via singlet energy transfer (step 1). It decays by photoinduced electron transfer step 2 to give the C-Pz-P$^{\bullet+}$-Q$^{\bullet-}$-Q initial charge-separated state. Two charge shift reactions operating in parallel (steps 3 and 4) generate two new charge-separated states C-Pz-P$^{\bullet+}$-Q-Q$^{\bullet-}$ and C-Pz$^{\bullet+}$-P-Q$^{\bullet-}$-Q. The C-Pz$^{\bullet+}$-P-Q$^{\bullet-}$-Q species can proceed by pathway 5, producing C-Pz$^{\bullet+}$-P-Q-Q$^{\bullet-}$, or pathway 6 to generate C$^{\bullet+}$-Pz-P-Q$^{\bullet-}$-Q. These two intermediate states can produce the final C$^{\bullet+}$-Pz-P-Q-Q$^{\bullet-}$ species via steps 8 or 9. Competing with all of these electron transfer steps are decay of the initial excited states by the usual photophysical pathways (steps 11 and 12) and charge recombination of each intermediate. The most rapid charge recombination step is undoubtedly step 10, as the radical ions are most strongly coupled at this stage of charge separation. The quantum yield of step 2 is 0.85, whereas the quantum yield of the final C$^{\bullet+}$-Pz-P-Q-Q$^{\bullet-}$ state is 0.83. Thus, all of the charge separation pathways compete successfully with charge recombination. This high yield is attributable not only to the proper tuning of the thermodynamic driving force and electronic coupling for each reaction, but also to the fact that the crucial initial charge recombination step 10 is competed against by two pathways operating in parallel (steps 3 and 4).

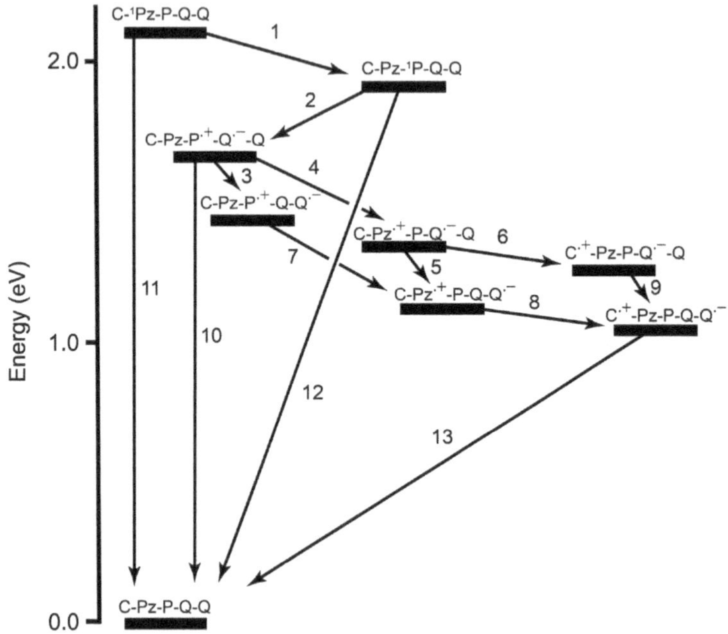

Figure 3 Transient states and relaxation pathways for pentad **10**. State energies are estimated from cyclic voltammetric data.

The lifetime of $C^{\bullet+}$-Pz-P-Q-$Q^{\bullet-}$ is 55 μs. The lifetime of the final state in the version of the pentad in which both porphyrins are present as the free base is about 200 μs in dichloromethane.

The pentad **10** illustrates that with careful molecular engineering that uses the theoretical concepts devised by Marcus and others to favour desired electron transfer events over competing processes, and by using both parallel and sequential electron transfers, it is possible to prepare artificial photosynthetic reaction centres that approach the natural ones in terms of quantum yield, fraction of photon energy stored, and lifetime of charge separation.

The Japanese group consisting of Mataga, Maruyama, Osuka, Okada and others was very active in the synthesis and study of large and complex multicomponent artificial reaction centres in the 1990s. An example is molecule **11** (Fig. 4), which consists of a stacked zinc porphyrin dimer (D) which is linked to a third zinc porphyrin (P) bearing a pyromellitimide (I)—quinone (Q) electron acceptor moiety (Osuka et al., 1995). Excitation of the porphyrin system is followed by multistep electron transfers that generate a $D^{\bullet+}$-P-I-$Q^{\bullet-}$ charge-separated state with a lifetime of 2.6 μs.

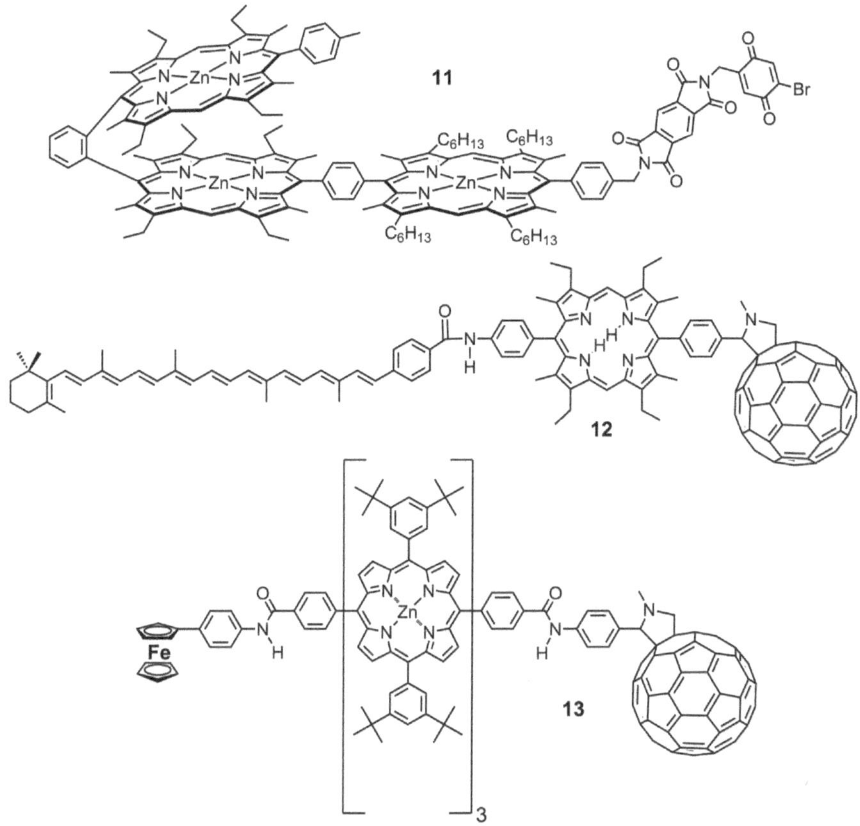

Figure 4 Five-component artificial reaction centre **11**, carotenoporphyrin-fullerene triad **12** and ferrocene-containing pentad **13**.

4.2.2 Fullerene-Based Artificial Reaction Centres

As mentioned earlier, in the mid-1990s, fullerenes replaced quinones as the favoured acceptors for electron transfer from porphyrin excited states. Thus multicomponent artificial reaction centres began to appear based on this motif. Surprisingly, some unexpected new behaviours appeared in these molecules. The first triad was carotenoporphyrin–fullerene **12** (Fig. 4) (Liddell et al., 1997). The molecule is clearly modelled on triad **7** and its relatives, but the quinone is replaced by a fullerene derivative (C_{60}). In 2-methyltetrahydrofuran solution, excitation of the porphyrin yields the first excited singlet state C-^{1}P-C_{60}, which decays by photoinduced electron and singlet energy transfer to the fullerene with a time constant τ of 10 ps. The fullerene singlet state also decays by photoinduced electron transfer ($\tau = 32$ ps). The product of photoinduced electron transfer from both C-^{1}P-C_{60} and C-P-$^1C_{60}$ is C-$P^{\bullet+}$-$C_{60}^{\bullet-}$, which is formed with a quantum yield of 0.99. At this stage, the molecule has converted light to electrochemical potential. However, based on studies of model dyads, charge recombination of C-$P^{\bullet+}$-$C_{60}^{\bullet-}$ to yield the ground state occurs with $\tau = 3.3$ ns. This lifetime is considerably longer than those achieved with dyads such as **3**, but is still so short that accessing the stored redox potential via chemical reactions would be challenging. In **12**, however, a rapid shift of the positive charge to the carotene competes with charge recombination, yielding $C^{\bullet+}$-P-$C_{60}^{\bullet-}$. The final charge-separated state has a lifetime of 170 ns and is formed with a quantum yield of 0.14. In benzonitrile, the lifetime is 770 ns and the yield is 0.12.

Interestingly, the major product of decay of $C^{\bullet+}$-P-$C_{60}^{\bullet-}$ is not the ground state, but rather the carotenoid triplet state, which decays to the ground state with a time constant of 4.9 μs. Charge recombination to form chlorophyll and carotenoid triplet states is also observed in natural photosynthetic preparations (Dutton, Leigh, & Seibert, 1972; Regev, Nechushatai, Levanon, & Thornber, 1989; Rutherford, Paterson, & Mullet, 1981; Thurnauer, Katz, & Norris, 1975), but is rare in artificial photosynthetic molecules (Hasharoni et al., 1995). The triplet state of **12** forms by the radical pair mechanism. The $C^{\bullet+}$-P-$C_{60}^{\bullet-}$ state is initially formed as a singlet biradical, which comes to equilibrium with the three essentially degenerate triple sublevels during the lifetime of charge separation. Recombination of the triplet biradical to the carotenoid spectroscopic triplet state is more rapid than recombination of the singlet biradical to the ground state.

Photoinduced electron transfer in **12** to yield $C^{\bullet+}$-P-$C_{60}^{\bullet-}$ occurs even at temperatures below 8 K (Kuciauskas, Liddell, Moore, Moore, & Gust, 1998). This behaviour contrasts drastically with that of similar triads based on quinone electron acceptors, and indeed most other artificial photosynthetic reaction centres. Quinone-based molecules such as **7** can undergo rapid photoinduced electron transfer in solution, but electron transfer typically ceases when the solvent is frozen to a glass. In the glass, the solvent molecules are no longer able to reorient to stabilize the newly formed charges. Thus, the energy of the charge-separated state increases to the point that photoinduced electron transfer is no longer possible for thermodynamic reasons, and the molecule decays by the usual photophysical pathways. Because of the low sensitivity of the fullerene radical anion and porphyrin radical cation to solvent stabilization relative to quinones, the loss of stabilization of the charge-separated state when the solvent freezes is no longer sufficient to prevent electron transfer. Interestingly, natural reaction centres also undergo photoinduced electron transfer down to around 4 K.

In bacterial photosynthesis, the time constant for recombination of the charge-separated state in the reaction centre to a triplet state is affected by relatively small external magnetic fields. The same is true in **12** (Kuciauskas et al., 1998). In a glass at 77 K, photoinduced electron transfer yields $C^{\bullet+}$-P-$C_{60}^{\bullet-}$, which decays on the microsecond time scale to yield ^{3}C-P-C_{60}. When a small (20 mT) magnetic field is applied, the lifetime of $C^{\bullet+}$-P-$C_{60}^{\bullet-}$ increases by about 50%. At zero field, the initially formed singlet biradical state is in equilibrium with the three triplet biradical sublevels, and all four states have comparable populations. Decay to the carotenoid triplet only occurs from the three triplet sublevels. In the presence of the field, the S and T_0 states are still rapidly interconverting, but the T_+ and T_- states are isolated from the other two due to the electronic Zeeman interaction, and are not significantly populated. Under these conditions, recombination to the triplet occurs only from T_0, and the lifetime of the charge-separated state increases. Studies on a closely related triad suggest that related magnetic field effects could be used by birds and other animals to orient and navigate in the earth's magnetic field. It was found that the lifetime of $C^{\bullet+}$-P-$C_{60}^{\bullet-}$ in this triad is altered by magnetic fields close to that of the earth's ($\leq$50 μT), and that this response is anisotropic, establishing the feasibility of chemical magnetoreception in organisms (Maeda et al., 2008).

Just as the first quinone-based triad prompted the preparation and study of numerous other triads and more complex molecules, the first fullerene-based triad sparked studies of other, even more complex molecules. One example of many is pentad **13**, which was reported by Imahori, Fukuzumi and coworkers (Imahori et al., 2004). This research group has contributed a very large amount of research in the artificial reaction centre area. Time-resolved spectroscopic studies show that excitation of ferrocene (Fc) zinc porphyrin fullerene **13** yields the porphyrin excited states, which decay to generate $Fc^{\bullet+}$-$(ZnP)_3$-$C_{60}^{\bullet-}$ after several electron transfer steps. This final state is formed with at quantum yield of 0.83 in benzonitrile, and at 163 K in N,N-dimethylformamide has a lifetime of 0.53 s. This molecule is an extreme example of the lifetimes and yields available with artificial reaction centres which can be achieved with both good design and significant expenditure of synthetic effort.

5. METAL COMPLEX–BASED ARTIFICIAL REACTION CENTRES

In the work described thus far, the emphasis has been on artificial reaction centres with porphyrins and their relatives as the primary light absorbers and electron donors. Parallel with the development of this field has been work in which transition metal complexes play these roles. Much of the research has employed ruthenium complexes with 2,2′-bipyridine ligands. In most of these systems, the chromophore forms a metal-to-ligand charge-transfer (MLCT) state upon absorption of visible light, and this state can act as either an electron donor to an attached acceptor moiety or an acceptor from an attached donor species. The group of T. J. Meyer has been a pioneer in this area (Meyer, 1989; Sullivan et al., 1978; Westmoreland, Lebozec, Murray, & Meyer, 1983). For example, complex **14** (Fig. 5) is functionally a donor–acceptor dyad in which the ruthenium complex is the light absorber and excited state donor, and a bound pyridinium group acts as an electron acceptor (Sullivan et al., 1978). Evidence for generation of a charge-separated state upon illumination was observed.

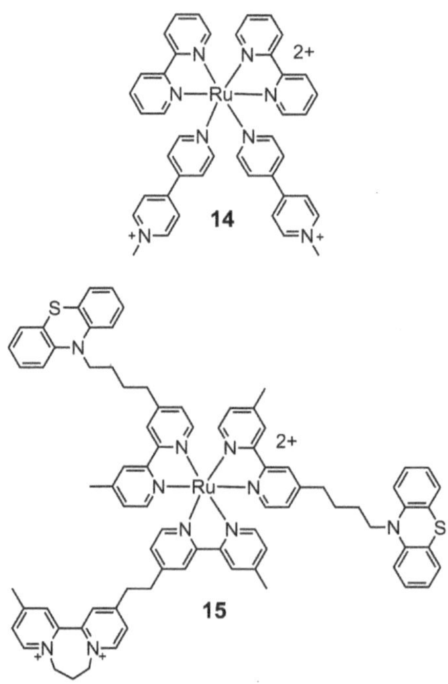

Figure 5 Ruthenium-based donor–acceptor dyad (**14**) and triad (**15**).

The first metal complex triad is molecule **15** (Fig. 5), which is composed of a ruthenium tris-bipyridyl derivative linked to a pyridinium moiety and two phenothiazines (Danielson, Elliott, Merkert, & Meyer, 1987). Upon excitation, the MLCT state either donates an electron to the pyridinium species or accepts an electron from a phenothiazine to yield an intermediate charge-separated state. In either intermediate, subsequent electron transfer yields a final state with a positive charge on the phenothiazine and reduced pyridinium moiety, returning the ruthenium chromophore to its original state. Thus, the molecule undergoes a two-step electron transfer reaction, giving a final charge-separated state with a lifetime of 165 ns in dichloromethane with a quantum yield of 0.26. It is an analogue of triads **7** and **8**.

As with the porphyrin systems, the report of triad **15** led to a flurry of research on metal complex donor—acceptor systems, including very complex molecules that can produce long-lived charge-separated states, and the topic has been reviewed extensively (Balzani, Juris, Venturi, Campagna, & Serroni, 1996; Scandola, Chiorboli, Indelli, & Rampi, 2001). The group of Balzani and coworkers centred at the University of Bologna has been especially active in this area, but many groups have made valuable contributions. For example, Sauvage, Balzani and coworkers reported a triad related to **16** involving a ruthenium bis-terpyridine chromophore which avoids some of the stereochemical ambiguity inherent in the tris-bipyridine complexes (Collin et al., 1991).

Complex **16** (Fig. 6) (Molnar, Nallas, Bridgewater, & Brewer, 1994) consists of three metal centres: an iridium complex covalently linked to

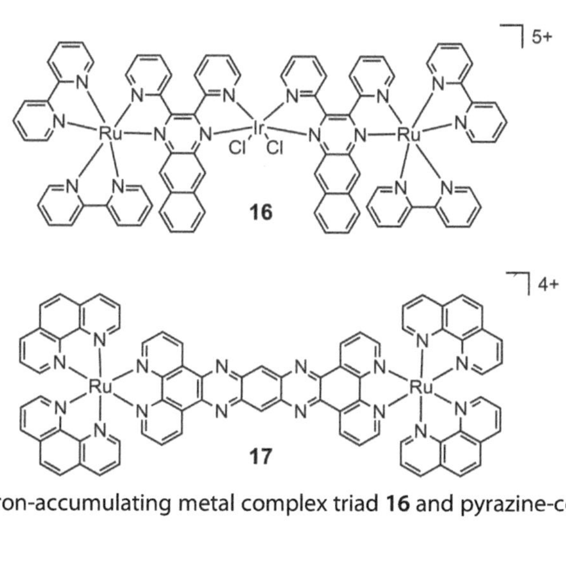

Figure 6 Electron-accumulating metal complex triad **16** and pyrazine-centred triad **17**.

two ruthenium centres. Light absorption by either ruthenium complex is followed by photoinduced electron transfer to the iridium moiety. If the oxidized ruthenium is then reduced by a sacrificial donor in solution, absorption of a second photon by a ruthenium centre results in donation of a second electron to the iridium complex. Thus, the iridium species collects two reducing equivalents. This idea is important, because if solar energy is to be used to produce fuels, multiple redox equivalents from successive absorption of photons must be stored to carry out multielectron chemistry such as water oxidation to oxygen and reduction of protons to hydrogen fuel.

Complex **17** (Fig. 6), reported by Scandola, Campagna and coworkers, features a pyrazine-containing bridge between two ruthenium chromophores (Chiorboli et al., 2003; Konduri et al., 2002). Excitation of a ruthenium moiety initiates electron transfer to the bridge to yield a charge-separated state with a very long lifetime of 1.3 μs in dichloromethane. In the presence of added sacrificial electron donors, **17** can accumulate up to four electrons.

6. ARTIFICIAL ANTENNAS

Most of the sunlight used for natural photosynthesis is not absorbed by reaction centres, but rather by antenna systems. The antennas consist of large arrays of chromophores that absorb light and transfer the resulting excitation energy through the antenna array and ultimately to the reaction centre. Antennas perform a number of useful functions. They deliver excitation energy to reaction centres rapidly enough to allow the multielectron redox processes necessary for water oxidation and fuel production to occur at a reasonable rate. They incorporate regulatory and photoprotective functions that reduce photodamage to the photosynthetic membranes in bright sunlight. Antennas also harvest light at wavelengths where chlorophylls do not absorb strongly. Due to these important functions, it seems likely that antennas could be important in artificial photosynthesis as well.

6.1 Porphyrin-Based Artificial Antennas

Antenna function depends on singlet–singlet energy transfer between chromophores, which in turn requires that the light absorbers be in close spatial proximity. One way to control interchromophore separations and orientations is to link them with chemical bonds. Thus, there has long been an interest in multiporphyrin arrays as models for photosynthetic antennas, and a large number have been reported. Only a very few illustrative examples will be given here. One early example is **18** (Fig. 7), which was reported in 1983 (Milgrom, 1983). Singlet–singlet energy transfer among the zinc porphyrins of **18** was inferred by the author, but not observed directly.

Structure **18** lacks some of the features of an ideal artificial antenna system. As the field developed, researchers constructed arrays with more rigid linkages between the chromophores, and shorter distances between adjacent units. These attributes help control and optimize some of the major factors that affect the efficiency of energy transfer. More recent antenna **19** (Fig. 8) is an array of six zinc porphyrins organized by a central hexaphenylbenzene core (Cho et al., 2003). The distances between porphyrin units are strictly controlled, although some rotation about single bonds in the linkages is possible. Time-resolved spectroscopic studies showed that with Q-band

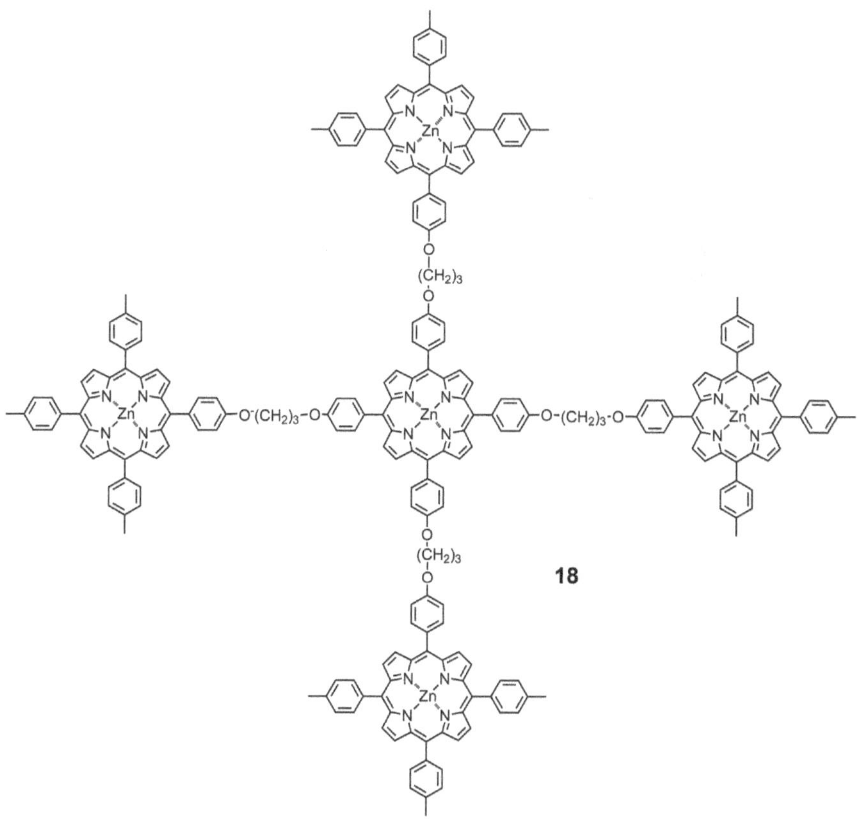

Figure 7 Early porphyrin array as an antenna mimic.

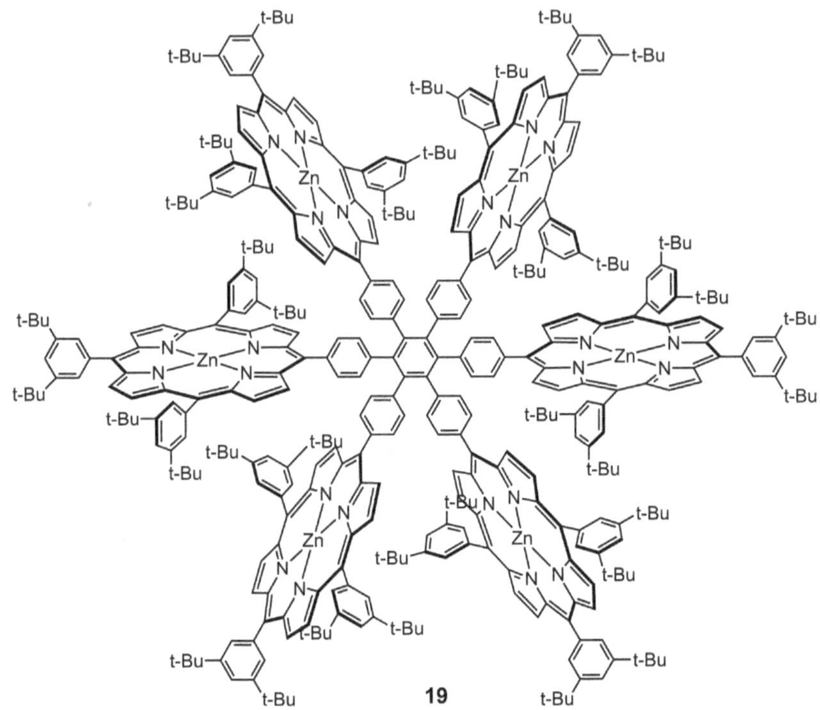

19

Figure 8 Wheel-shaped porphyrin hexad artificial antenna.

excitation, energy transfer between adjacent porphyrins occurs with a time constant of 180 ps.

As mentioned earlier, natural antenna systems often use chromophores other than chlorophyll to assure adequate coverage of the solar spectrum. Building on this theme, molecule **20** (Fig. 9) uses a hexaphenylbenzene core to organize three different kinds of antenna units which together cover the solar spectrum from the ultraviolet to around 600 nm (Terazono et al., 2009). The borondipyrromethene chromophores absorb in the 330–430 nm range, and the bis(phenylethynyl)anthracene units absorb in the 400–500 nm region. Transient spectroscopic experiments revealed that these units transfer excitation energy to the porphyrins, which absorb out to c. 600 nm, by both direct and multistep pathways as shown in Fig. 9 to generate the porphyrin excited singlet state in nearly quantitative yield.

Unlike hexad **19**, heptad **20** not only absorbs light and transfers excitation efficiently among the chromophores, but also includes an artificial reaction centre in the form of the porphyrin–fullerene unit. The excitation energy arriving at a porphyrin initiates photoinduced electron transfer to the

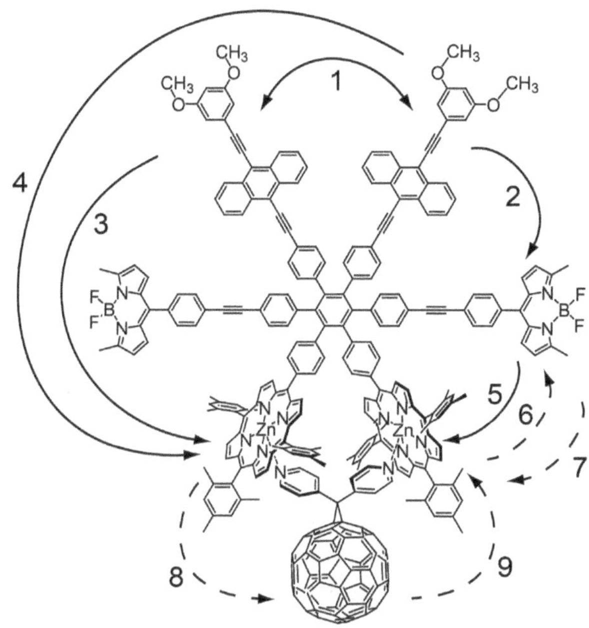

20

Figure 9 Artificial photosynthetic multiantenna-reaction centre **20**.

fullerene with a quantum yield of essentially one to generate a charge-separated state, which has a lifetime of 230 ps in 1,2-difluorobenzene.

There are many examples of linear and dendritic porphyrin arrays that could be used as artificial antennas. In recent years, some extremely long linear systems have been reported, especially by the groups of Osuka (Aratani, Kim, & Osuka, 2009), Anderson (Taylor & Anderson, 1999) and Therien (Duncan, Susumu, Sinks, & Therien, 2006; Lin, Dimagno, & Therien, 1994).

6.2 Carotenoid-Based Artificial Antennas

Carotenoid polyenes occur in all natural photosynthetic organisms, where they carry out several functions. Antennas that use carotenoids to absorb light around 450 nm where chlorophyll is not efficient are extremely important, harvesting a good fraction of the light used worldwide for photosynthesis. The photophysics of carotenoids suggests that they would be difficult to employ in this way. Direct absorption into their first excited singlet state S_1 is symmetry forbidden, and not observed. The well-known orange colour of carotenes arises from the strongly allowed S_2 transition. The S_2 state has a very short lifetime (typically hundreds of femtoseconds), and relaxes to the S_1 state, which is also short lived (a few picoseconds) and decays to the ground state. Because it is a forbidden state, fluorescence from the S_1 state is not typically observed. The forbidden nature and short lifetime of the carotenoid S_1 state mean that transfer of excitation from this state to another chromophore, such as a chlorophyll or porphyrin, will require strong interaction between the chromophores, and therefore very close spatial approach. Although the S_2 state is strongly allowed, its extremely short lifetime means that energy transfer involving that state also requires strong interactions. In spite of these difficulties, carotenoids are very efficient antennas in natural photosynthesis.

Designing and synthesizing artificial molecules in which carotenoids can act as antennas for porphyrins or other chromophores is therefore much more challenging than carrying out these tasks with porphyrins or many other chromophores. However, given the importance of carotenes in photosynthesis, researchers have been investigating model systems for some time. The first covalently linked carotenoporphyrin, **21** (Fig. 10), was reported in 1980 (Dirks, Moore, Moore, & Gust, 1980). In spite of the short linkage between the *para* position of a *meso*-phenyl ring on the porphyrin macrocycle and the carotenoid, no energy transfer from the carotene to the porphyrin was observed. However, when the carotenoid linkage was moved to the *ortho* position, as in **22**, an energy transfer efficiency of ~25% was found (Moore, Dirks, Gust, & Moore, 1980). In **21**, the carotenoid extends away from the plane of the porphyrin macrocycle, and the π-electron systems of the two chromophores do not interact sufficiently well that energy transfer can compete with decay of the carotenoid excited states by internal conversion to the ground state. In **22**, nuclear magnetic resonance studies show that the carotene is stacked above the porphyrin macrocycle at a distance of 4–5 Å, and this very close approach of the two π-electron systems allows energy transfer to compete with the other relaxation pathways available to the carotenoid excited states.

Research has continued in this area, and much has been learnt about the complex photochemistry of carotenoids in both natural and artificial photosynthesis. This knowledge has permitted the design of molecules in which the carotenoid is a much more efficient antenna. For example, carotenophthalocyanine **23** (Fig. 10) consists of a carotenoid with nine carbon–carbon double bonds linked to a zinc phthalocyanine via an amide bridge. Fluorescence excitation and time-resolved spectroscopic studies show that the overall efficiency of singlet energy transfer from the carotenoid to the phthalocyanine is about 90%. The transient studies reveal that the efficiency of transfer from the carotenoid S_2 state to the phthalocyanine is about 69%, whereas that from the S_1

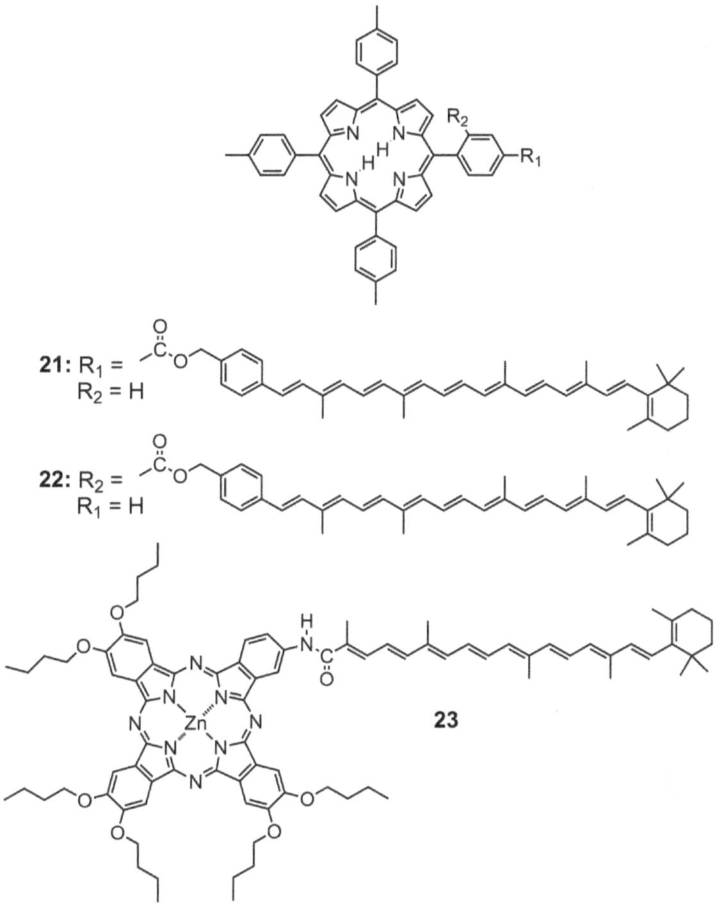

Figure 10 Isomeric carotenoporphyrins **21** and **22** and carotenophthalocyanine **23**.

state is about 20% (Berera et al., 2007). In general, it is found that the details of energy transfer between carotenoids and cyclic tetrapyrroles depends strongly on both the structures of the chromophores (especially the number of double bonds in the carotenoid) and the nature of the chemical linkage between them.

7. PHOTOPROTECTION

Although human society in its present form is made possible by the availability of various sources of technological energy, humans are well aware of the dangers of dealing with high-energy sources. Fires, explosions and electrical shocks occur when technological energy gets out of control. Photosynthetic organisms, and especially those doing aerobic photosynthesis, face similar dangers. Visible and ultraviolet photon energy is high, relative to the energy of chemical bonds, and the production of reactive oxygen species during photosynthesis can harm or kill the organism. Consequently, photosynthetic organisms have evolved control mechanisms, photoprotective systems and repair mechanisms to limit such damage. It seems likely that if organic or metal–organic molecules are to be used for artificial photosynthesis, they will need photoprotection as well. Although most of the research into artificial photosynthesis has gone into developing the photoinduced charge separation process and subsequent catalytic steps, some work has been done on photoprotection. Indeed some of this research was done relatively early in the evolution of artificial photosynthesis. Two examples will be discussed.

7.1 Carotenoid Photoprotection Against Singlet Oxygen

Excited state singlet oxygen ($O_2(a^1\Delta_g)$) is highly reactive with many organic compounds, including the lipids of photosynthetic membranes. It can be produced by the reaction of the usual ground-state triple oxygen with the triplet excited states of chlorophylls, porphyrins and other chromophores via an energy transfer process. For example, in the reaction centres of purple bacteria, the normal electron transport chain can be blocked by reduction of the quinones, in which case the charge-separated states formed by photoinduced electron transfer cannot evolve, and recombine instead to yield bacteriochlorophyll triplet states. These states are potent sensitizers of singlet oxygen. The bacteria are protected from photodamage by singlet oxygen through the presence of carotenoid polyenes in the reaction centre. The carotenoids rapidly quench bacteriochlorophyll triplet states by a triplet–triplet energy transfer process to yield the carotenoid triplet state (Frank, Machnicki, & Friesner, 1983; Schenck, Mathis, & Lutz, 1984; Takiff & Boxer, 1988). The carotenoid triplet is of low energy and has a short lifetime: it decays harmlessly to the ground state, evolving heat.

Carotenoids have been known to quench singlet oxygen in solution since 1968 (Foote & Denny, 1968), but preventing singlet oxygen sensitization by diffusive quenching of the sensitizer triplet state is limited by the rate of diffusion, and only observed at very high concentrations of carotenoids. Carotenoporphyrin dyads, in which the two chromophores are covalently linked, avoid this limitation. The photosynthetic carotenoid photoprotective mechanism was first mimicked in a model compound in 1981

(Bensasson et al., 1981) using carotenoporphyrins **21** and **22** (Fig. 10). Fig. 11 shows the rate of destruction of the dye diphenylisobenzofuran (DPBF) in aerated toluene solution when the solution is illuminated in the presence of various sensitizers and carotenoids (Bensasson et al., 1981). If the solution contains a tetra-arylporphyrin sensitizer alone, singlet oxygen is produced under illumination, and rapidly reacts with DPBF, resulting in the bleaching of the dye (hollow circles in Fig. 11). If either of two carotenoids are added to the solution of dye and sensitizer (hollow squares and triangles), the rate of destruction of the dye is only slightly decreased because at the concentrations employed ($\sim 2 \times 10^{-6}$ M), diffusion-controlled triplet energy transfer from the sensitizer triplet state to carotene cannot compete with quenching of the sensitizer triplet by oxygen to generate singlet oxygen. When the carotene is covalently linked to the porphyrin as in molecule **21**, a significant decrease in the rate of photodestruction of the dye is observed because triplet—triplet energy transfer from the porphyrin triplet state to the carotenoid is occurring sufficiently rapidly to reduce the triplet lifetime of the porphyrin to the point that reaction with O_2 is no longer efficient. In molecule **22**, which features a folded arrangement of the porphyrin and the carotenoid as discussed earlier,

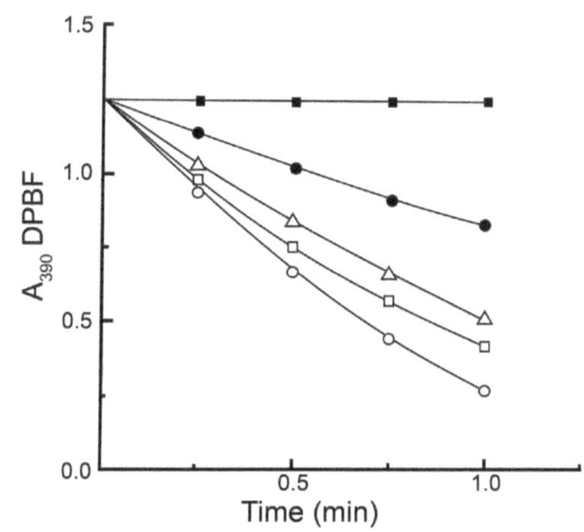

Figure 11 Bleaching of the absorbance of the dye diphenylisobenzofuran by singlet oxygen as a function of illumination time. The dye was dissolved in aerated toluene containing equimolar amounts of either a porphyrin sensitizer (*hollow circles*), the porphyrin sensitizer plus 7′-apo-7′-(4-hydroxymethylphenyl)-β-carotene (*hollow squares*), the porphyrin sensitizer plus β-carotene (*hollow triangles*), carotenoporphyrin **21** (*filled circles*) or carotenoporphyrin **22** (*filled squares*). Note the complete photoprotection provided by **22**.

photoprotection is essentially complete due to very rapid quenching of the porphyrin triplet state. Time-resolved measurements show that this energy transfer occurs on the time scale of a few nanoseconds.

Subsequent research has probed the role of intramolecular motions in control of the triplet energy transfer rate (Gust et al., 1985, 1992; Moore et al., 1982), the mimicry of the tetrapyrrole-mediated relay of triplet energy from chlorophyll to carotenoids in photosynthesis (Gust, Moore, Moore, Krasnovsky, et al., 1993, 1998) and the details of the triplet transfer process using ultrafast spectroscopic techniques (Berera et al., 2006; Kloz et al., 2011; Kodis et al., 2004).

7.2 Light-Dependent Regulation of the Artificial Photosynthetic Process

In general, under conditions of bright sunlight the antenna systems of photosynthetic organisms collect far more light than can be used by the reaction centre. While this situation allows the organism to survive at low light intensities, at high light it results in the photogeneration of oxidation and reduction equivalents more rapidly than in the dark, catalytic reactions of photosynthesis can use them to make fuels. The result is photodamage due to these high-energy intermediates and their reaction products. Photosynthetic organisms have developed various mechanisms for regulating the photosynthetic process as a function of light intensity. One of these mechanisms is called nonphotochemical quenching (NPQ). The process takes different forms in different organisms. In cyanobacteria, the phycobilisome antenna system is associated with a soluble carotenoid protein called orange carotenoid protein that is involved in the NPQ process (Wilson et al., 2006, 2008). Under low light conditions, this protein is orange in colour and has no effect on the photosynthetic process. Under high light, the protein converts to red carotenoid protein, which quenches the excitation collected by the antenna before it reaches the reaction centre. Thus, the red form down-regulates photosynthesis. The red form thermally reverts to the inactive orange form. Thus, the amount of protein in the red form increases as the light intensity increases, and decreases as the light intensity decreases. The net result is that the quantum yield of the photosynthetic process is reduced as the light intensity increases, and increases as the light intensity decreases.

In 2008, this general photosynthetic regulatory process was mimicked using molecular pentad **24** (Fig. 12) (Straight et al., 2008). Molecule **24** consists of a porphyrin—fullerene artificial reaction centre much like those discussed earlier that bears two bis-(phenylethynyl)anthracene antennas

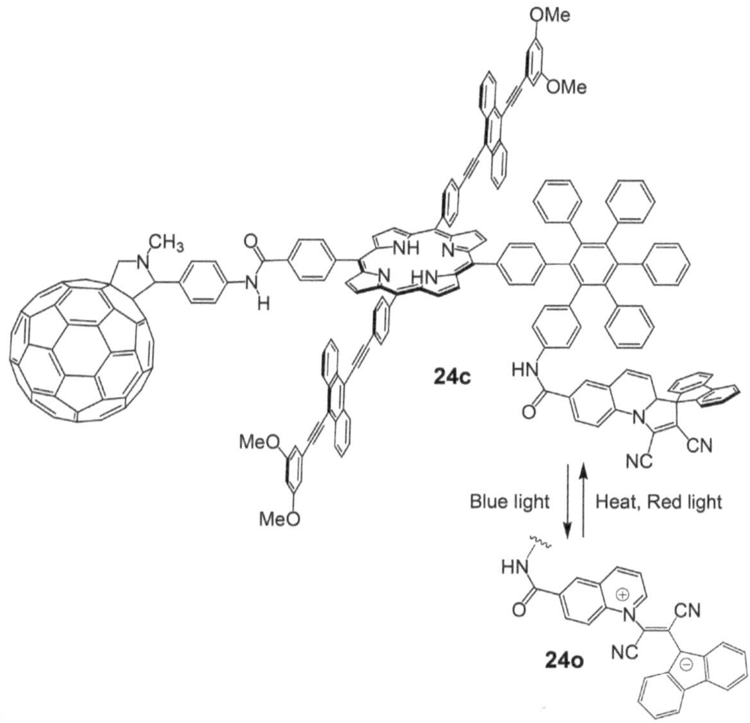

Figure 12 Molecular pentad that functionally mimics the nonphotochemical quenching process in cyanobacteria.

and a photochromic dye. In the dark or under low light conditions, the dye exists in the dihydroindolazine (DHI, **24c**) form, which is colourless in the visible and has no effect on the photochemistry of the rest of the molecule. When DHI absorbs light, it photoisomerizes into a coloured betaine (BT), as in **24o**. Molecule **24o** thermally converts back to **24c** with a time constant of 37 s at 25°C.

Spectroscopic studies of **24** indicate that when the porphyrin moiety of **24c** is excited, it donates an electron to the fullerene with a time constant of 2.4 ns to form a charge-separated state with a quantum yield of 0.82. Light absorbed by the bis-(phenylethynyl)anthracene is transferred to the porphyrin with a quantum yield of unity, and also leads to photoinduced electron transfer. Thus, **24c** acts as an efficient artificial photosynthetic antenna-reaction centre combination. When **24c** is isomerized to **24o**, the resulting BT efficiently quenches the porphyrin first excited singlet state by energy transfer with a time constant of only 33 ps, thus essentially

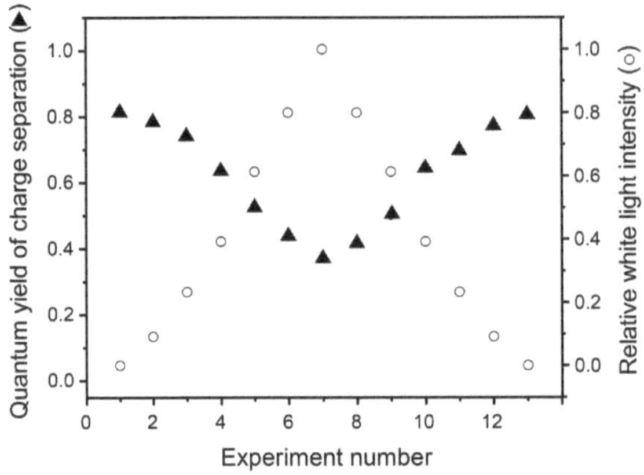

Figure 13 The net quantum yield of photoinduced electron transfer in **24** (*solid triangles*) is shown for 13 experiments in which the intensity of white light irradiating the sample (*open circles*) was varied. At low light intensity, the quantum yield of charge separation is 0.82 whereas for the highest light intensity, the yield was only 0.37.

precluding any photoinduced electron transfer. The resulting BT excited state decays to the ground state with the liberation of heat.

This behaviour allows **24** to mimic NPQ in cyanobacteria. At low light levels, a sample of the compound in solution exists mainly as **24c** and undergoes photoinduced electron transfer with a quantum yield of 0.82. As the light intensity is raised, increasing amounts of the **24c** are converted to **24o**, which has a quantum yield of charge separation of only 1%, and the overall yield of charge separation decreases. When the light intensity is lowered again, the overall yield increases. This behaviour is shown in Fig. 13, which shows the results of 13 experiments in which a sample of **24** was illuminated with white light of various intensities. At low light, the quantum yield photoinduced electron transfer was 0.82, but at the highest intensities investigated, it was reduced to 0.37. Lowering the light level returned the quantum yield to its initial value. This behaviour is functionally analogous to the way that orange carotenoid protein controls the yield of photosynthesis in cyanobacteria.

8. HARVESTING THE ENERGY OF CHARGE SEPARATION – SOLAR FUEL PRODUCTION

The culmination of artificial photosynthesis is the use of the redox equivalents generated by artificial reaction centres to produce a chemical fuel. Making a fuel is complex because in addition to the reaction centre and any associated antenna, the process requires catalysts and construction of a complete system in which the reaction centre and catalysts can work together. This is a difficult problem, but some progress has been made, and the rate of discovery is accelerating at the present time. Here, I will give a few examples of how fuel production can be accomplished. None of these systems is as yet viable for practical fuel production, but they illustrate that the entire fuel generation process is achievable.

8.1 ATP, a Biological Fuel

As was mentioned in Section 4, the ultimate electron acceptor in the photosynthetic reaction centre is a quinone molecule. The quinone is reduced to a semiquinone, and finally to a hydroquinone after two photoinduced electron transfers. This reduction involves the uptake of two protons from water on the cytoplasmic side of the membrane. In bacterial photosynthesis, the hydroquinone diffuses to the next component of the apparatus, a proton pump called the cytochrome bc_1 complex. This complex oxidizes the hydroquinone back to a quinone, using the energy released to translocate protons across the membrane and establish a proton concentration and charge imbalance (proton motive force, pmf). The oxidation process is ultimately driven, via various cytochrome redox relays, by the oxidized special pair, which becomes reduced to its initial state. Finally, the enzyme ATP synthase allows protons to flow back across the membrane, down the thermodynamic gradient, driving the release of ATP (adenosine triphosphate) formed from adenosine diphosphate and inorganic phosphate. The ATP is an energy-rich molecule that fills the majority of the energy needs of the bacterium. Thus, ATP is a fuel that powers the organism, and its generation requires energy stored as a transmembrane pmf.

An analogue of this process has been demonstrated in the laboratory (Steinberg-Yfrach et al., 1997). The artificial transmembrane proton pump comprises carotene—porphyrin—quinone triad **25** and lipid-soluble quinone **26** (Fig. 14). These molecules are housed in the lipid bilayer membrane of a liposome vesicle with a diameter of ~150 nm (Fig. 15). The triad is inserted vectorially into the membrane so that for the majority of the molecules the quinone moiety, which bears a negatively charged carboxylate group at the pH of the solution, remains near the hydrophilic exterior. Quinone **26**, on the other hand, is soluble only in the hydrophobic membrane interior, and is more easily reduced than the naphthoquinone moiety of the triad.

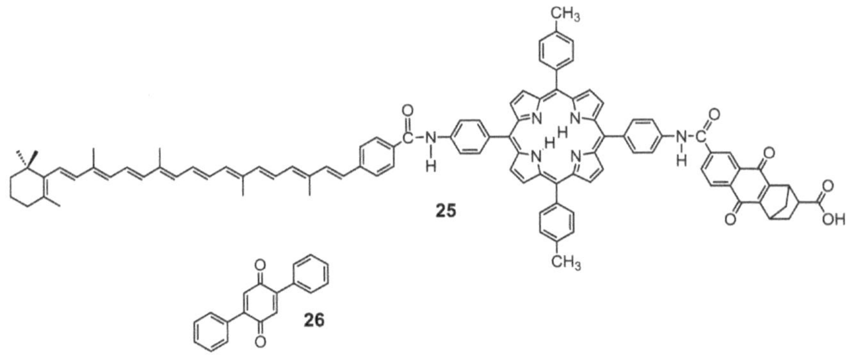

Figure 14 Components of the transmembrane proton pump.

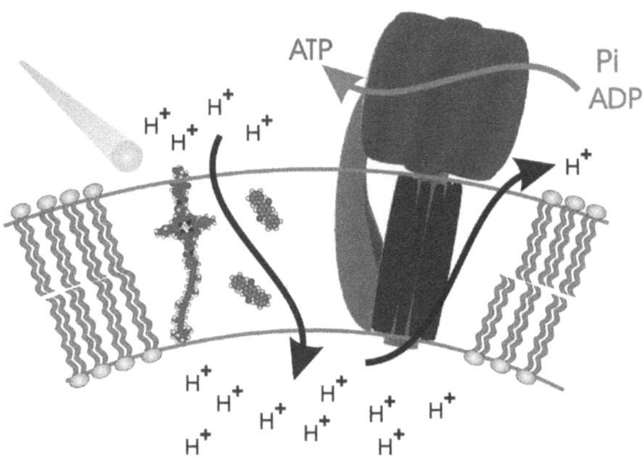

Figure 15 Artificial photosynthetic biological power plant that converts light energy into chemical energy stored in ATP.

The pump is based on a redox loop (Mitchell, 1966). It is powered by triad **25** which, upon absorption of light by the porphyrin, generates a $C^{\bullet+}$-P-$Q^{\bullet-}$ charge-separated state as discussed in Section 4.1. Although all details of the function of the pump have not been elucidated, the following description illustrates the major aspects of the process. The quinone radical anion of $C^{\bullet+}$-P-$Q^{\bullet-}$ near the outer membrane surface reduces a molecule of quinone **26** to yield the semiquinone anion, which is basic enough to accept a proton from the exterior aqueous environment. The resulting neutral semiquinone radical diffuses within the bilayer. When it encounters the carotenoid radical cation near the inner membrane surface, it is oxidized back to the quinone. The protonated quinone is a strong acid (pK_a c. −6) that releases the proton into the inner volume of the liposome. The net result is proton translocation into the vesicle interior and regeneration of the photoredox catalyst. The redox loop-based proton pump can in principle be driven from various combinations of redox levels of quinone **26**.

The action of this pump acidifies the solution inside the liposome, and this can be detected by inclusion of a suitable pH-sensitive dye. A maximum ΔpH of ~2 units was obtained. Because the proton pumping photocycle does not translocate any compensating charge, a transmembrane electrical potential is also established. This was detected by the use of a suitable fluorescent dye. Both the proton gradient and the potential gradient comprise the energy stored by the system (pmf).

As mentioned earlier, photosynthetic bacteria use the transmembrane pmf to power the synthesis of ATP. The energy generated by the flow of protons through enzyme ATP synthase from the low-pH side of the membrane to the high-pH side is used to produce the ATP. This enzyme was coupled with the artificial proton pump to produce a complete system for the generation of ATP using light energy (Fig. 15) (Steinberg-Yfrach et al., 1998). The enzyme was incorporated vectorially into the liposomal membrane, and the proton pump system was added so that the proton pumping photocycle translocated hydrogen ions into the liposome, and the enzyme transported them back out of the enzyme, using the pmf to synthesize ATP from adenosine diphosphate and inorganic phosphate. The process was followed using the firefly luciferin—luciferase system, in which the consumption of one molecule of ATP results in emission of one photon, which could be detected in the 570-nm region. More than 100 ATP per second per molecule of ATP were produced under light saturating conditions.

8.2 Production of Technologically Useful Fuels

Although ATP is a ubiquitous fuel for organisms, it is useless as a fuel for technology. The emphasis from the beginnings of artificial photosynthesis research until the present has been the production of useful fuels such as hydrogen gas or a reduced carbon fuel. In the last few years, artificial photosynthesis research has reached the degree of sophistication necessary to allow construction of simple but complete systems for production of fuel using sunlight. Although this research is in its infancy, the progress to date is encouraging. A few examples of complete solar fuel systems that feature at least one molecular active component are discussed later.

A solar fuel production system must include a chemical species that can be oxidized to serve as a source of electrons, another species that can be reduced to a useful fuel by these electrons, a reaction centre analogue that can use sunlight to carry out photoinduced electron transfer to generate oxidation and reduction equivalents and catalysts for the oxidation and reduction processes. All components must be integrated into a single functional device. Most of the initial experiments in artificial photosynthetic fuel production have been directed towards oxidation of water to produce hydrogen ions, molecular oxygen and electrons, and reduction of the hydrogen ions to hydrogen gas, which is an attractive fuel. In 2008, Mallouk and coworkers reported the photoelectrochemical cell diagrammed schematically in Fig. 16. The cell has two electrodes immersed in an aqueous buffer solution. The photoanode is adapted from the nanoparticulate titanium dioxide wide band gap semiconductor electrodes developed for dye-sensitized solar cells that produce electricity. The nanoparticles are coated onto a conductive, transparent glass substrate and sensitized with a ruthenium tris-bipyridyl—type metal complex. This complex in turn is linked to nanoparticulate iridium oxide, which functions as a water oxidation catalyst. The second electrode, which is linked to the photoanode via a wire, is metallic platinum. Excitation of the ruthenium complex with light is

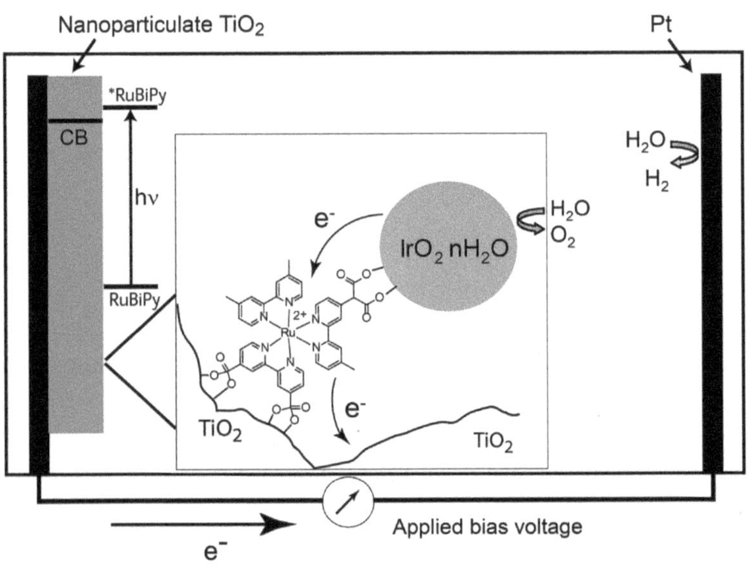

Figure 16 Complete photoelectrochemical cell for solar splitting of water to hydrogen and oxygen.

followed by electron injection into the nanoparticulate TiO_2 electrode, from which electrons move into the wire and migrate to the Pt cathode, where they reduce hydrogen ions to hydrogen gas. The oxidized ruthenium complex is reduced by the attached iridium oxide nanoparticle. Accumulation of at least four oxidizing equivalents on the iridium oxide is followed by oxidation of water to oxygen gas and hydrogen ions. Thus, the cell uses light energy to split water.

Although this cell functions and demonstrates the possibility of solar fuel production by artificial photosynthesis, it requires a small electrical bias in order to work at a reasonable rate, and the quantum yield of hydrogen is very low (about 1%). The quantum yield has been improved to 2.3% by inclusion of a molecular relay to transport electrons from the iridium oxide to the oxidized ruthenium dye (Megiatto et al., 2012). This relay is based on the principles of proton-coupled electron transfer that are important in the natural photosynthetic system for transporting electrons from the water oxidation complex of photosystem II to the oxidized chlorophyll of the reaction centre. In more recent work, porphyrins have been found to function in place of the ruthenium dye on the photoanode (Swierk et al., 2015).

A more efficient photochemical water-splitting cell has been reported recently by Sun and coworkers (Gao et al., 2013). The overall design is similar to that in Fig. 16. However, the water oxidation catalyst is ruthenium complex **27** (Fig. 17) after it has been attached to the TiO_2 photoanode (with concomitant loss of the ethoxy groups). Under illumination and after application of a bias of 0.2 V, a photocurrent density of more than 1.7 mA/cm^2 was obtained.

A cell featuring a novel approach to nanoparticulate electrode preparation and a molecular combination light absorber and water oxidation catalyst was recently reported by Meyer and coworkers (Alibabaei et al., 2013). The cell is similar to that in Fig. 16, but the photoanode employs indium tin oxide nanoparticles covered with a thin shell of titanium dioxide. The sensitizer dye is a ruthenium tris-bipyridyl derivative that is covalently attached to a ruthenium-based water oxidation catalyst. The structure, **28**, is given in Fig. 17. At peak photocurrent, this cell features a photon conversion efficiency of 4.4%.

The theoretical efficiency of a water-splitting cell using a single dye is limited to about 30% by thermodynamic considerations (Shockley & Queisser, 1961). By using tandem configurations in which two or more cells, each with a different sensitizer dye, are used in series, higher efficiencies should be possible. In these cells, two or more photons must be absorbed per

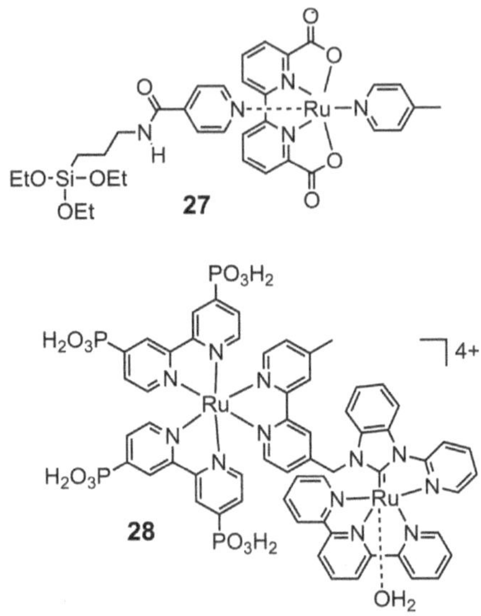

Figure 17 A molecular catalyst for water oxidation in an artificial photosynthetic solar water splitting cell (**27**) and a combination sensitizer and water oxidation catalyst (**28**).

electron used in redox reactions, but at least twice as much of the solar spectrum becomes available for use while still achieving adequate overpotentials for water oxidation. Such tandem cells are being investigated for solar water splitting (Alibabaei et al., 2013; Fan et al., 2014; Wood et al., 2014). For example, Graetzel, Sivula and coworkers have reported such a cell that uses a metal oxide photoanode in tandem with a dye-sensitized solar cell and achieves an efficiency of 3.1% (Brillet et al., 2012).

9. CONCLUSIONS

This brief overview demonstrates that great strides have been made in artificial photosynthesis research over the last 30 years. In the late 1970s the theoretical basis for photoinduced electron transfer was not fully understood/appreciated, the structure and details of natural photosynthesis had not been elucidated at the molecular level and the theoretical, synthetic and analytical tools for designing, synthesizing and studying large artificial photosynthetic molecular constructs and similar systems had not been developed. Today, although artificial photosynthesis has certainly not evolved to the state that practical solar fuel production is possible, the basic outlines of such a process have been demonstrated. Rapid progress is being made, and if research in solar fuels continues at this pace, the next few years should at least make clear which approaches are likely to be viable. Although it is not yet clear whether a molecule-based (or indeed any other) approach will be practically useful, it must be remembered that the only proven system for solar fuel production is photosynthesis, which is based on organic molecules and metal—organic complexes. A replacement for fossil fuels must be identified eventually if human society is to continue to prosper. As Ciamician said in his 1912 paper, 'If our black and nervous civilization, based on coal, shall be followed by a quieter civilization based on the utilization of solar energy, that will not be harmful to progress and to human happiness' (Ciamician, 1912).

REFERENCES

Alibabaei, L., Brennaman, M., Norris, M. R., Kalanyan, B., Song, W., Losego, M. D.... Meyer, T. J. (2013). Solar water splitting in a molecular photoelectrochemical cell. *Proceedings of the National Academy of Sciences of the United States of America, 110*, 20008—20013.

Alibabaei, L., Luo, H., House, R. L., Hoertz, P. G., Lopez, R., & Meyer, T. J. (2013). Applications of metal oxide materials in dye sensitized photoelectrosynthesis cells for making solar fuels: let the molecules do the work. *Journal of Materials Chemistry A, 1*, 4133—4145.

Aratani, N., Kim, D., & Osuka, A. (2009). Discrete cyclic porphyrin arrays as artificial light-harvesting antenna. *Accounts of Chemical Research, 42*, 1922—1934.

Asahi, T., Ohkouchi, M., Matsusaka, R., Mataga, N., Zhang, R. P., Osuka, A., & Maruyama, K. (1993). Intramolecular photoinduced charge separation and charge recombination of the product ion pair states of a series of fixed-distance dyads of porphyrins and quinones: energy gap and temperature dependences of the rate constants. *Journal of the American Chemical Society, 115*, 5665—5674.

Balzani, V., Juris, A., Venturi, M., Campagna, S., & Serroni, S. (1996). Luminescent and redox-active polynuclear transition metal complexes. *Chemical Reviews, 96*, 759—833.

Baran, P. S., Monaco, R. R., Kahn, A. U., Schuster, D. I., & Wilson, S. R. (1997). Synthesis and cation-mediated electronic interactions of two novel classes of porphyrin-fullerene hybrids. *Journal of the American Chemical Society, 119*, 8363—8364.

Bell, T. D. M., Smith, T. A., Ghiggino, K. P., Ranasinghe, M. G., Shephard, M. J., & Paddon-Row, M. N. (1997). Long-lived photoinduced charge separation in a bridged C_{60}-porphyrin dyad. *Chemical Physics Letters, 268*, 223—228.

Bensasson, R. V., Land, E. J., Moore, A. L., Crouch, R. L., Dirks, G., Moore, T. A., & Gust, D. (1981). Mimicry of antenna and photoprotective carotenoid functions by a synthetic carotenoporphyrin. *Nature, 290*, 329—332.

Berera, R., Herrero, C., van Stokkum, L. H. M., Vengris, M., Kodis, G., Palacios, R. E.... Kennis, J. T. M. (2006). A simple artificial light-harvesting dyad as a model for excess energy dissipation in oxygenic photosynthesis. *Proceedings of the National Academy of Sciences of the United States of America, 103*, 5343—5348.

Berera, R., van Stokkum, I. H. M., Kodis, G., Keirstead, A. E., Pillai, S., Herrero, C.... Kennis, J. T. M. (2007). Energy transfer, excited-state deactivation, and exciplex formation in artificial caroteno-phthalocyanine light-harvesting antennas. *Journal of Physical Chemistry B, 111*, 6868—6877.

Brillet, J., Yum, J. H., Cornuz, M., Hisatomi, T., Solarska, R., Augustynski, J. ... Sivula, K. (2012). Highly efficient water splitting by a dual-absorber tandem cell. *Nature Photonics, 6*, 823–827.
Calcaterra, L., Closs, G. L., & Miller, J. R. (1983). Fast intramolecular electron transfer in radical ions over long distances across rigid saturated hydrocarbon spacers. *Journal of the American Chemical Society, 105*, 670–671.
Calvin, M. (1978). Simulating photosynthetic quantum conversion. *Accounts of Chemical Research, 11*, 369–374.
Chiorboli, C., Fracasso, S., Scandola, F., Campagna, S., Serroni, S., Konduri, R., & MacDonnell, F. M. (2003). Primary charge separation in photoinduced multielectron storage systems. A dinuclear ruthenium(II) species featuring a charge-separated state with a lifetime of 1.3 μs. *Chemical Communications*, 1658–1659.
Cho, H. S., Rhee, H., Song, J. K., Min, C.-K., Takase, M., Aratani, N. ... Kim, D. (2003). Excitation energy transport processes of porphyrin monomer, dimer, cyclic trimer, and hexamer probed by ultrafast fluorescence anisotropy decay. *Journal of the American Chemical Society, 125*, 5849–5860.
Ciamician, G. (1912). The photochemistry of the future. *Science, 36*, 385–394.
Closs, G. L., & Miller, J. R. (1988). Intramolecular long-distance electron transfer in organic molecules. *Science, 240*, 440–446.
Collin, J. P., Guillerez, S., Sauvage, J. P., Barigelletti, F., Decola, L., Flamigni, L., & Balzani, V. (1991). Photoinduced processes in dyads and triads containing a ruthenium(II) bis(terpyridine) photosensitizer covalently linked to electron-donor and acceptor groups. *Inorganic Chemistry, 30*, 4230–4238.
Connolly, J. S., & Bolton, J. R. (1988). Intramolecular electron transfer: history and some implications for artificial photosynthesis. In M. A. Fox, & M. Chanon (Eds.), *Photoinduced electron transfer, part D* (pp. 303–393). Amsterdam: Elsevier.
Danielson, E., Elliott, C. M., Merkert, J. W., & Meyer, T. J. (1987). Photochemically induced charge separation at the molecular level. A chromophore quencher complex containing both an electron transfer donor and an acceptor. *Journal of the American Chemical Society, 109*, 2519–2520.
DeGraziano, J. M., Liddell, P. A., Leggett, L., Moore, A. L., Moore, T. A., & Gust, D. (1994). Free energy dependence of photoinduced charge separation rates in porphyrin dyads. *Journal of Physical Chemistry, 98*, 1758–1761.
Deisenhofer, J., Epp, O., Miki, K., Huber, R., & Michel, H. (1985). Structure of the protein subunits in the photosynthetic reaction centre of *Rhodopseudomonas viridis* at 3A resolution. *Nature, 318*, 618–624.
Deisenhofer, J., Epp, O., Sinning, I., & Michel, H. (1995). Crystallographic refinement at 2.3-Angstrom resolution and refined model of the photosynthetic reaction-center from *Rhodopseudomonas viridis*. *Journal of Molecular Biology, 246*, 429–457.
Dirks, G., Moore, A. L., Moore, T. A., & Gust, D. (1980). Light absorption and energy transfer in polyene-porphyrin esters. *Photochemistry and Photobiology, 32*, 277–280.
Drovetskaya, T., Reed, C. A., & Boyd, P. D. W. (1995). A fullerene porphyrin conjugate. *Tetrahedron Letters, 36*, 7971–7974.
Duncan, T. V., Susumu, K., Sinks, L. E., & Therien, M. J. (2006). Exceptional near-infrared fluorescence quantum yields and excited-state absorptivity of highly conjugated porphyrin arrays. *Journal of the American Chemical Society, 128*, 9000–9001.
Dutton, P. L., Leigh, J. S., & Seibert, M. (1972). Primary processes in photosynthesis: in situ ESR studies on the light induced oxidized and triplet state of reaction center bacteriochlorophyll. *Biochemical and Biophysical Research Communications, 46*, 406–413.
Fan, K., Li, F., Wang, L., Daniel, Q., Gabrielsson, E., & Sun, L. (2014). Pt-free tandem molecular photoelectrochemical cells for water splitting driven by visible light. *Physical Chemistry Chemical Physics, 16*, 25234–25240.
Foote, C. S., & Denny, R. W. (1968). *Journal of the American Chemical Society, 90*, 6233.

Frank, H. A., Machnicki, J., & Friesner, R. A. (1983). Energy transfer between the primary donor bacteriochlorophyll and carotenoids in *Rhodopseudomonas sphaeroides*. *Photochemistry and Photobiology, 38*, 451–455.

Fujishima, A., & Honda, K. (1972). Electrochemical photolysis of water at a semiconductor electrode. *Nature, 238*, 37–38.

Gao, Y., Ding, X., Liu, J., Wang, L., Lu, Z., Li, L. ... Sun, L. (2013). Visible light driven water splitting in a molecular device with unprecedentedly high photocurrent density. *Journal of the American Chemical Society, 135*, 4219–4222.

Guldi, D. M., & Asmus, K.-D. (1997). Electron transfer from C76 (C2v') and C78 (D2) to radical cations of various arenes: evidence for the Marcus inverted region. *Journal of the American Chemical Society, 119*, 5744–5745.

Gust, D., Mathis, P., Moore, A. L., Liddell, P. A., Nemeth, G. A., Lehman, W. R. ... Chachaty, C. (1983). Energy transfer and charge separation in carotenoporphyrins. *Photochemistry and Photobiology, 37S*, S46.

Gust, D., & Moore, T. A. (2000). Intramolecular photoinduced electron transfer reactions of porphyrins. In K. M. Kadish, K. M. Smith, & R. Guilard (Eds.), *The porphyrin handbook* (Vol. 8, pp. 153–190). New York: Academic Press.

Gust, D., Moore, T. A., Bensasson, R. V., Mathis, P., Land, E. J., Chachaty, C. ... Nemeth, G. A. (1985). Stereodynamics of intramolecular triplet energy transfer in carotenoporphyrins. *Journal of the American Chemical Society, 107*, 3631–3640.

Gust, D., Moore, T. A., Moore, A. L., Devadoss, C., Liddell, P. A., Hermant, R. M. ... Gouni, I. (1992). Triplet and singlet energy transfer in carotene-porphyrin dyads: role of the linkage bonds. *Journal of the American Chemical Society, 114*, 3590–3603.

Gust, D., Moore, T. A., Moore, A. L., Krasnovsky, A. A., Jr., Liddell, P. A., Nicodem, D. ... Pessiki, P. J. (1993). Mimicking the photosynthetic triplet energy transfer relay. *Journal of the American Chemical Society, 115*, 5684–5691.

Gust, D., Moore, T. A., Moore, A. L., Kuciauskas, D., Liddell, P. A., & Halbert, B. D. (1998). Mimicry of carotenoid photoprotection in artificial photosynthetic reaction centers: triplet-triplet energy transfer by a relay mechanism. *Journal of Photochemistry and Photobiology, B, 43*, 209–216.

Gust, D., Moore, T. A., Moore, A. L., Lee, S.-J., Bittersmann, E., Luttrull, D. K. ... Trier, T. T. (1990). Efficient multistep photoinduced electron transfer in a molecular pentad. *Science, 248*, 199–201.

Gust, D., Moore, T. A., Moore, A. L., Macpherson, A. N., Lopez, A., DeGraziano, J. M. ... Kerrigan, P. K. (1993). Photoinduced electron and energy transfer in molecular pentads. *Journal of the American Chemical Society, 115*, 11141–11152.

Harrison, R. J., Pearce, B., Beddard, G. S., Cowan, J. A., & Sanders, J. K. M. (1987). Photoinduced electron transfer in pyromellitimide-bridged porphyrins. *Chemical Physics, 116*, 429–448.

Hasharoni, K., Levanon, H., Greenfield, S. R., Gosztola, D. J., Svec, W. A., & Wasielewski, M. R. (1995). Mimicry of the radical pair and triplet states in photosynthetic reaction centers with a synthetic model. *Journal of the American Chemical Society, 117*, 8055–8056.

Hush, N. S. (1958). Adiabatic rate processes at electrodes. I. Energy-charge relation. *Journal of Chemical Physics, 28*, 962–972.

Hush, N. S. (1961). Adiabatic theory of outer-sphere electron-transfer reactions in solutions. *Transactions of the Faraday Society, 57*, 557–580.

Imahori, H., Hagiwara, K., Akiyama, T., Aoki, M., Taniguchi, S., Okada, T. ... Sakata, Y. (1996). The small reorganization energy of C_{60} in electron transfer. *Chemical Physics Letters, 263*, 545–550.

Imahori, H., Hagiwara, K., Akiyama, T., Taniguchi, S., Okada, T., & Sakata, Y. (1995). Synthesis and photophysical property of porphyrin-linked fullerene. *Chemistry Letters*, 265–266.

Imahori, H., Sekiguchi, Y., Kashiwagi, Y., Sato, T., Araki, Y., Ito, O.... Fukuzumi, S. (2004). Long-lived charge-separated state generated in a ferrocene-meso, meso-linked porphyrin trimer-fullerene pentad with a high quantum yield. *Chemistry – A European Journal, 10*, 3184–3196.

Joran, A. D., Leland, B. A., Felker, P. M., Zewail, A. H., Hopfield, J. J., & Dervan, P. B. (1987). Effect of exothermicity on electron transfer rates in photosynthetic molecular models. *Nature, 327*, 508–511.

Joran, A. D., Leland, B. A., Geller, G. G., Hopfield, J. J., & Dervan, P. B. (1984). Models for photochemical electron transfer at fixed distances. Porphyrin-bicyclo[2.2.2]octane-quinone and porphyrin-bibicyclo[2.2.2]octane-quinone. *Journal of the American Chemical Society, 106*, 6090–6092.

Khundkar, L. R., Perry, J. W., Hanson, J. E., & Dervan, P. B. (1994). Weak temperature dependence of electron transfer rates in fixed-distance porphyrin-quinone model systems. *Journal of the American Chemical Society, 116*, 9700–9709.

Kloz, M., Pillai, S., Kodis, G., Gust, D., Moore, T. A., Moore, A. L.... Kennis, J. T. M. (2011). Carotenoid photoprotection in artificial photosynthetic antennas. *Journal of the American Chemical Society, 133*, 7007–7015.

Kodis, G., Herrero, C., Palacios, R., Mariño-Ochoa, E., Gould, S. L., de la Garza, L.... Kennis, J. T. M. (2004). Light harvesting and photoprotective functions of carotenoids in compact artificial photosynthetic antenna designs. *Journal of Physical Chemistry B, 108*, 414–425.

Konduri, R., Ye, H. W., MacDonnell, F. M., Serroni, S., Campagna, S., & Rajeshwar, K. (2002). Ruthenium photocatalysts capable of reversibly storing up to four electrons in a single acceptor ligand: a step closer to artificial photosynthesis. *Angewandte Chemie International Edition, 41*, 3185.

Kong, J. L., & Loach, P. A. (1978). In P. L. Dutton, & H. Scarpa (Eds.), *Frontiers of biological energetics: From electrons to tissues* (pp. 73–82). New York: Academic Press.

Kuciauskas, D., Liddell, P. A., Hung, S.-C., Lin, S., Stone, S., Seely, G. R.... Gust, D. (1997). Structural effects on photoinduced electron transfer in carotenoid-porphyrin-quinone triads. *Journal of Physical Chemistry, 101*, 429–440.

Kuciauskas, D., Liddell, P. A., Moore, A. L., Moore, T. A., & Gust, D. (1998). Magnetic switching of charge separation lifetimes in artificial photosynthetic reaction centers. *Journal of the American Chemical Society, 120*, 10880–10886.

Leland, B. A., Joran, A. D., Felker, P. M., Hopfield, J. J., Zewail, A. H., & Dervan, P. B. (1985). Picosecond fluorescence studies on intramolecular photochemical electron transfer in porphyrins linked to quinones at two different fixed distances. *Journal of Physical Chemistry, 89*, 5571–5573.

Levich, V. (1966). Present state of the theory of oxidation-reduction in solution (bulk and electrode reactions). *Advances in Electrochemistry and Electrochemical Engineering, 4*, 249–371.

Liddell, P. A., Kuciauskas, D., Sumida, J. P., Nash, B., Nguyen, D., Moore, A. L.... Gust, D. (1997). Photoinduced charge separation and charge recombination to a triplet state in a carotene-porphyrin-fullerene triad. *Journal of the American Chemical Society, 119*, 1400–1405.

Liddell, P. A., Sumida, J. P., Macpherson, A. N., Noss, L., Seely, G. R., Clark, K. N.... Gust, D. (1994). Preparation and photophysical studies of porphyrin-C_{60} dyads. *Photochemistry and Photobiology, 60*, 537–541.

Lin, V. S. Y., Dimagno, S. G., & Therien, M. J. (1994). Highly conjugated, acetylenyl bridged porphyrins - new models for light-harvesting antenna systems. *Science, 264*, 1105–1111.

Maeda, K., Henbest, K. B., Cintolelsi, F., Kuprov, I., Rodgers, C. T., Liddell, P. A.... Hore, P. J. (2008). Chemical compass model of avian magnetoreception. *Nature, 453*, 387–390.

Marcus, R. A. (1956). The theory of oxidation-reduction reactions involving electron transfer. *Journal of Chemical Physics, 24*, 966–978.
Marcus, R. A., & Sutin, N. (1985). Electron transfers in chemistry and biology. *Biochimica et Biophysica Acta, 811*, 265–322.
Megiatto, J. D., Antoniuk-Pablant, A., Sherman, B. D., Kodis, G., Gervaldo, M., Moore, T. A. … Gust, D. (2012). Mimicking the electron transfer chain in photosystem II with a molecular triad thermodynamically capable of water oxidation. *Proceedings of the National Academy of Sciences of the United States of America, 109*, 15578–15583.
Meyer, T. J. (1989). Chemical approaches to artificial photosynthesis. *Accounts of Chemical Research, 22*, 163–170.
Milgrom, L. R. (1983). Synthesis of some new tetra-arylporphyrins for studies in solar energy conversion. *Journal of the Chemical Society, Perkin Transactions, 1*, 2535–2539.
Miller, J. R., Beitz, J. V., & Huddleston, R. K. (1984). Effect of free energy on rates of electron transfer between molecules. *Journal of the American Chemical Society, 106*, 5057–5068.
Mitchell, P. (1966). Chemiosmotic coupling in oxidative and photosynthetic phosphorylation. *Biological Reviews of the Cambridge Philosophical Society, 41*, 445–502.
Molnar, S. M., Nallas, G., Bridgewater, J. S., & Brewer, K. J. (1994). Photoinitiated electron collection in a mixed-metal trimetallic complex of the form ([(Bpy)(2)Ru(Dpb)](2) Ircl2)(Pf6)(5) (Bpy=2,2′-Bipyridine and Dpb=2,3-Bis(2-Pyridyl)Benzoquinoxaline). *Journal of the American Chemical Society, 116*, 5206–5210.
Moore, A. L., Dirks, G., Gust, D., & Moore, T. A. (1980). Energy transfer from carotenoid polyenes to porphyrins. *Photochemistry and Photobiology, 32*, 691–696.
Moore, A. L., Joy, A. M., Tom, R., Gust, D., Moore, T. A., Bensasson, R. V., & Land, E. J. (1982). Photoprotection by carotenoids during photosynthesis: motional dependence of intramolecular energy transfer. *Science, 216*, 982–984.
Moore, T. A., Gust, D., Mathis, P., Mialocq, J. C., Chachaty, C., Bensasson, R. V. … Moore, A. L. (1984). Photodriven charge separation in a carotenoporphyrin quinone triad. *Nature, 307*, 630–632.
Nishitani, S., Kurata, N., Sakata, Y., Misumi, S., Karen, A., Okada, T., & Mataga, N. (1983). A new model for the study of multistep electron transfer in photosynthesis. *Journal of the American Chemical Society, 105*, 7771–7772.
Osuka, A., Yamada, H., Maruyama, K., Ohno, T., Nozaki, K., Okada, T. … Mataga, N. (1995). Charge separation in zinc diporphyrin-zinc porphyrin-pyromellitimide-quinone tetrads. *Chemistry Letters*, 591–592.
Osuka, A., Zhang, R. P., Maruyama, K., Yamazaki, I., & Nishimura, Y. (1992). Synthesis and intramolecular electron transfer reaction of distance-fixed quinone-linked porphyrins. *Bulletin of the Chemical Society of Japan, 65*, 2807–2813.
Regev, A., Nechushatai, R., Levanon, H., & Thornber, J. P. (1989). *Journal of Physical Chemistry, 93*, 2421–2426.
Rutherford, A. W., Paterson, D. R., & Mullet, J. E. (1981). A light-induced spin-polarized triplet detected by EPR in photosystem II reaction centers. *Biochimica et Biophysica Acta, 635*, 205–214.
Sakata, Y., Kishimoto, M., Nishitani, S., Tatemitsu, H., Misumi, S., Karen, A. … Gust, D. (1987). Conformational dependence of electron transfer in quinone-linked porphyrin. *Studies in Organic Chemistry, 31*, 427–432.
Scandola, F., Chiorboli, C., Indelli, M. T., & Rampi, M. A. (2001). Covalently linked systems containing metal complexes. In V. Balzani (Ed.), *Electron transfer in chemistry* (Vol. 3, pp. 337–408). Weinheim: Wiley-VCH.
Schenck, C. C., Mathis, P., & Lutz, M. (1984). Triplet formation and triplet decay in reaction centers from the photosynthetic bacterium *Rhodopseudomonas sphaeroides*. *Photochemistry and Photobiology, 39*, 407–417.

Sessler, J. L., Johnson, M. R., Creager, S. E., Fettinger, J. C., & Ibers, J. A. (1990). Synthesis and characterization of quinone-substituted octaalkyl porphyrin monomers and dimers. *Journal of the American Chemical Society, 112*, 9310–9329.

Shockley, W., & Queisser, H. J. (1961). Detailed balance limit of efficiency of P-N Junction solar cells. *Journal of Applied Physics, 32*, 510.

Steinberg-Yfrach, G., Liddell, P. A., Hung, S.-C., Moore, A. L., Gust, D., & Moore, T. A. (1997). Artificial photosynthetic reaction centers in liposomes: photochemical generation of transmembrane proton potential. *Nature, 385*, 239–241.

Steinberg-Yfrach, G., Rigaud, J.-L., Durantini, E. N., Moore, A. L., Gust, D., & Moore, T. A. (1998). Light-driven production of ATP catalyzed by F_0F_1-ATP synthase in an artificial photosynthetic membrane. *Nature, 392*, 479–482.

Straight, S. D., Kodis, G., Terazono, Y., Hambourger, M., Moore, T. A., Moore, A. L., & Gust, D. (2008). Self-regulation of photoinduced electron transfer by a molecular nonlinear transducer. *Nature Nanotechnology, 3*, 280–283.

Sullivan, B. P., Abruna, H., Finklea, H. O., Salmon, D. J., Nagle, J. K., Meyer, T. J., & Sprintschnik, H. (1978). Multiple emissions from charge-transfer excited-states of ruthenium(II)-polypyridine complexes. *Chemical Physics Letters, 58*, 389–393.

Swierk, J. R., Mendez-Hernandez, D. D., McCool, N. S., Liddell, P. A., Terazono, Y., Pahk, I. … Mallouk, T. E. (2015). Metal-free organic sensitizers for use in water-splitting dye-sensitized photoelectrochemical cells. *Proceedings of the National Academy of Sciences of the United States of America, 112*, 1681–1686.

Tabushi, I., Koga, N., & Yanagita, M. (1979). *Tetrahedron Letters, 257*.

Takiff, L., & Boxer, S. G. (1988). Phosphorescence spectra of bacteriochlorophylls. *Journal of the American Chemical Society, 110*, 4425–4426.

Taylor, P. N., & Anderson, H. L. (1999). Cooperative self-assembly of double-strand conjugated porphyrin ladders. *Journal of the American Chemical Society, 121*, 11538–11545.

Terazono, Y., Kodis, G., Liddell, P. A., Garg, V., Moore, T. A., Moore, A. L., & Gust, D. (2009). Multiantenna artificial photosynthetic reaction center complex. *Journal of Physical Chemistry B, 113*, 7147–7155.

Thurnauer, M. C., Katz, J. J., & Norris, J. R. (1975). The triplet state in bacterial photosynthesis. Possible mechanisms of the primary photo-act. *Proceedings of the National Academy of Sciences of the United States of America, 72*, 3270–3274.

Wasielewski, M. R., Niemczyk, M. P., Svec, W. A., & Pewitt, E. B. (1985). High-quantum-yield long lived charge separation in a photosynthetic reaction center model. *Journal of the American Chemical Society, 107*, 5562–5563.

Weber, E. J. (1989). *Common life in the middle ages*. Boston, MA: WGBH Educational Foundation.

Westmoreland, T. D., Lebozec, H., Murray, R. W., & Meyer, T. J. (1983). Multiple-state emission and intramolecular electron-transfer quenching in rhenium(I) bipyridine based chromophore quencher complexes. *Journal of the American Chemical Society, 105*, 5952–5954.

Wilson, A., Ajlani, G., Verbavatz, J. M., Vass, I., Kerfeld, C. A., & Kirilovsky, D. (2006). A soluble carotenoid protein involved in phycobilisome-related energy dissipation in cyanobacteria. *Plant Cell, 18*, 992–1007.

Wilson, A., Punginelli, C., Gall, A., Bonetti, C., Alexandre, M., Routaboul, J. M. … Kirilovsky, D. (2008). A photoactive carotenoid protein acting as light intensity sensor. *Proceedings of the National Academy of Sciences of the United States of America, 105*, 12075–12080.

Wood, C. J., Cheng, M., Clark, C. A., Horvath, R., Clark, I. P., Hamilton, M. L. … Gibson, E. A. (2014). Red-absorbing cationic acceptor dyes for photocathodes in tandem solar cells. *Journal of Physical Chemistry C, 118*, 16536–16546.

CHAPTER TWO

'Direct Conversion': Artificial Photosynthesis With Cyanobacteria

Q. Chen[a], D. Montesarchio[a], K.J. Hellingwerf[1]
University of Amsterdam, Amsterdam, The Netherlands
[1]Corresponding author: Email: K.J.Hellingwerf@uva.nl

Contents

Abstract

Cyanobacteria, the only bacteria that can carry out oxygenic, (ie, plant-type) photosynthesis, can be engineered with the methods of synthetic biology so that they acquire the ability to convert CO_2 directly into biofuel and/or commodity chemicals. In such an approach one bypasses the formation of the entire complex set of (macro)molecules that jointly form biomass. This approach has become known as 'direct conversion' (of CO_2 and sunlight into biofuel) and has been shown to be feasible for several products already, even upon significant scaleup. Here we explain this concept of 'direct conversion' through natural photosynthesis and discuss its limitations and potential further improvement.

List of Abbreviations

C(3) C(3)-type oxygenic photosynthesis
C(4) C(4)-type oxygenic photosynthesis
PAR Photosynthetically active radiation
PSI Photosystem I
PSII Photosystem II

[a] These authors contributed equally to this manuscript.

Advances in Botanical Research, Volume 79
ISSN 0065-2296
http://dx.doi.org/10.1016/bs.abr.2016.03.001

1. INTRODUCTION

In order to not obstruct a development towards a sustainable future, our society will have to make a fundamental transition from an economy that derives the main part of its energy from the net flow, and oxidation, of carbon from fossil sources into atmospheric CO_2, into an economy in which the global carbon cycle will have been transformed into a closed (ie, circular form) driven principally by solar energy. Many adaptations will be necessary for such a transition to become feasible, including an overall reduction in energy consumption/dissipation, increased energy efficiency of many existing activities/technologies and the large-scale use of renewable electricity. The latter form of energy may take a primary role in future society and can and must be generated with techniques directly or indirectly tapping from solar energy, for example, with photovoltaic panels, hydroelectric- and tidal power stations and wind turbines. In addition to electricity, the solar energy can also be directly converted into chemical (free) energy, with the use of so-called 'artificial leaves' (Nocera, 2012) in compounds like H_2, CO and CH_3OH. This approach, however, has not yet resulted in large-scale demo plants, nor has it allowed the direct synthesis of molecules containing one or more carbon—carbon bonds.

Next to these man-made processes of harvesting solar energy, there is also an additional, natural, mechanism to achieve the same tapping of solar energy for human use: photosynthesis. This process is carried out by living organisms, ranging from the smallest microorganisms to plants, and is principally aimed at producing free energy to drive the synthesis/formation of new cells. For this synthesis, next to the basic building blocks and minerals required for biomass formation, free energy is also required in two forms: redox intermediates such as NAD(P)H and phosphorylated intermediates such as ATP (Hellingwerf & Konings, 1985), with the chemiosmotic free energy of a proton gradient as an exchangeable intermediate (Mitchell, 1957). Generally, and particularly during autotrophic growth (ie, with CO_2 as the carbon source), the building blocks for the synthesis of new cells will be more oxidized than the average redox level of biomass, although there are well-known exceptions to this rule when, for example, reduced sugars like mannitol are fermented or when photoheterotrophic bacteria grow with butyrate, for example, as their carbon source (see following sections). But we restrict the discussion here to photosynthesis, and more particularly to photoautotrophy.

Surprisingly, and so far little appreciated, four different types of photosynthesis have developed during evolution (Fig. 1): Generally accepted terminology has not yet been crystallized for these different types, but mostly they are referred to as: retinal-based photosynthesis (A), anoxygenic photosynthesis (B, C) and oxygenic photosynthesis (D). For the sum of (B) to (D) also the term chloro-photosynthesis or chloro-phototrophy is being used (eg, Kim et al., 2015). In the field of microbiology the existence and discovery of retinal-based photosynthesis has blurred the distinction between chemotrophs and phototrophs because sequence analysis only of

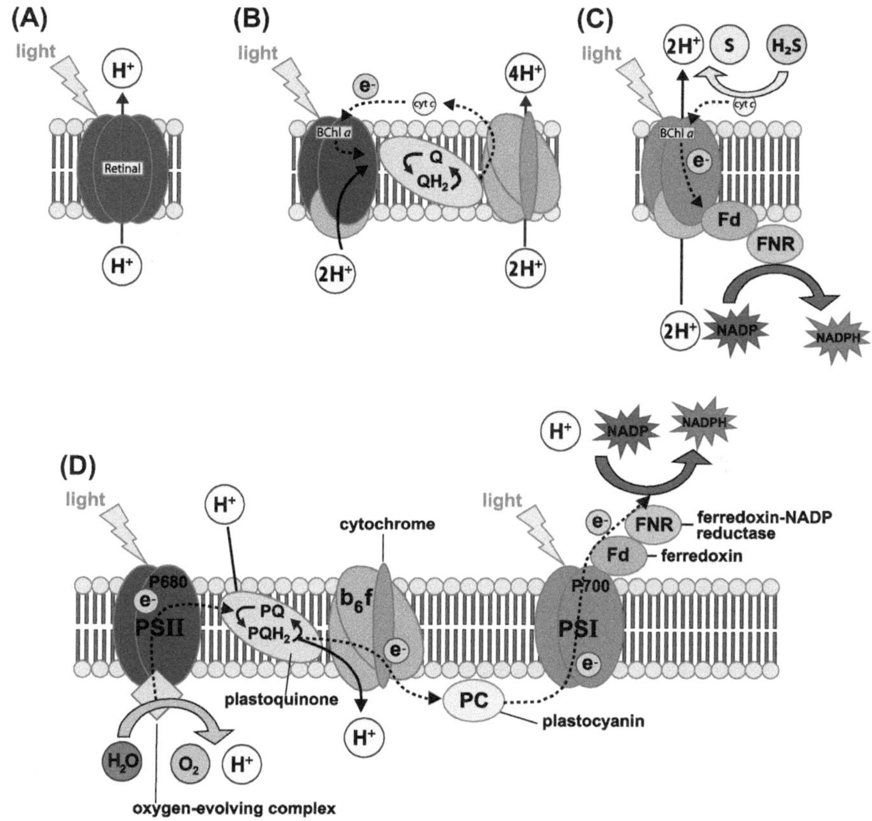

Figure 1 ***Four different types of photosynthesis: (A) Retinal photosynthesis, (B,C) Anoxygenic photosynthesis and (D) Oxygenic photosynthesis.*** The main photosynthetic complexes that operate in the four different types of photosynthesis are indicated as embedded in a bilayer membrane, connected with mobile electron/proton carriers. Also reference is made to the build-up of a proton motive force, as a result of the light-induced reactions (*E/Z* isomerization in (A) and electron transfer in (B—D)).

a newly discovered strain is not enough to decide between these two categories because of the existence of retinal proteins (ie, rhodopsins) with either a sensing function or a function in proton pumping (ie, photosynthetic energy transduction).

In photosynthesis of types A and B, light energy is converted into the free energy of hydrolysis of ATP only, via formation of a proton gradient as the high intermediate. The capacity to form NAD(P)H in these organisms is limited as this has to proceed via the so-called 'reversed electron transfer', driven also by the energy from the proton gradient (Mitchell, 1957). In contrast, both in anoxygenic photosynthesis of type C and in oxygenic photosynthesis, light energy is directly converted not only into ATP, but also into redox carriers such as NADPH. This is because the electrons activated/energized by photon energy are transported downhill through a linear pathway of electron-transfer components that eventually end in the formation of a compound that is much more reduced than the respective electron donor for the process. It is also this latter aspect that distinguishes the two types of photosynthesis; for anoxygenic photosynthesis, electron donors like various sulphur compounds and manganese and ferrous salts, that is, electron donors with a moderate redox midpoint potential and a significant environmental abundance, are available.

Mechanistically there is a very large difference between retinal-based photosynthesis and (bacterio)chlorophyll-based photosynthesis (or 'chlorophotosynthesis' (Kim et al., 2015)). The former process is based on transient energy storage in the reconfiguration of a C=C double bond of a retinal chromophore, catalysed by a single protein that makes very limited use of antennae pigments (Danon & Stoeckenius, 1974; Oesterhelt & Stoeckenius, 1971). The latter makes use of a complex protein-based machinery that facilitates light-driven electron transfer in one or two reaction centres and makes use of very extensive antenna systems that channel excitons to the centres (Berg, John, & Lubert, 2012). At the global scale, chlorophotosynthesis is by far the most important, but estimates have been made indicating that retinal-based photosynthesis may contribute up to 10% of solar energy conversion (Ebrey, 2002).

2. OXYGENIC PHOTOSYNTHESIS

The moderate redox midpoint potential of the electron donors that are suitable to facilitate anoxygenic photosynthesis (Fig. 1C) also assures that a single light reaction suffices to bring an electron to a redox level so that it can be transferred spontaneously to NAD(P)H. In this latter aspect lies the principal difference with oxygenic photosynthesis. In this type of photosynthesis, water (abundantly available in many ecosystems), with its very high redox midpoint potential, can be used as the electron donor for NAD(P)H formation, which, however, because of the large redox-span, necessitates the involvement of two successively operating photosystems, that is, II (PSII) and I (PSI), that together with the connecting redox components plastoquinone, the cytochrome b_6/f complex and plastocyanine, form the Z-scheme of oxygenic- or plant-type photosynthesis (Fig. 1D).

Illumination of the components of the Z-scheme embedded in the thylakoid membranes of plants and microorganisms accordingly gives rise to the formation of NADPH and ATP, such that, ideally, with the input of eight photons, two molecules of NADPH and three molecules of ATP are formed. Stress conditions, however, may alter this ratio (further discussed in Section 4). This 2:3 stoichiometry exactly matches the requirements of the metabolic pathway that functions to channel the main carbon-source building block of photosynthesis, that is, CO_2, into new (metabolic intermediates for) cell material/biomass. This metabolic pathway is known as the Calvin–Benson cycle (Berg et al., 2012) and involves as the key enzyme 'the most abundant protein on earth', RubisCO, that catalyses the incorporation of CO_2 into ribulose-1,5-bis-phosphate, to form two molecules of 3-phosphoglycerate. Accordingly, for the synthesis of one molecule of phosphoglycerate 24 photons will be necessary and its formation is accompanied by the uptake of three molecules of CO_2 and an equimolar release of oxygen. The 3-phosphoglycerate can then be converted into any of the complex set of biomolecules necessary to make a new cell, primarily via the lower part of glycolysis in combination with the tricarboxylic acid cycle, to form fatty- and amino acids, and the gluconeogenic pathway to form sugars.

Oxygenic photosynthesis has evolved to become the dominant type of photosynthesis, both in the terrestrial environment (via plants) and in the aquatic environment (via macro- and microalgae). In cyanobacteria, this process is characterized by a relative overabundance of PSI (as compared to PSII), the occurrence of multiple forms of chlorophyll (eg, a, d, f) and phycobilisome antennae that fill the gap between the two main absorption bands of chlorophyll. Their photosynthesis is very efficient, among others because of the presence of carboxysomes, which largely prevents photorespiration by RubisCO, but their energy metabolism is complex because of the strongly interactive nature of respiratory and photosynthetic electron flow in their interconnected thylakoid and cytoplasmic membranes (Madigan et al., 2014).

Retinal-based photosynthesis and anoxygenic photosynthesis have an important role only in selected ecological niches. These niches can be differentiated with respect to the specific electron donor that is available, other than water. If very little to none is available, that is, in oligotrophic environments such as the open oceans or the surface of glaciers, organisms carrying out retinal photosynthesis may thrive (Akram et al., 2013). In eutrophied waters, in which a significant amount of fixed carbon is present, presumably mostly as organic acids, anoxygenic photosynthesis of type that makes use of quinone electron acceptors (displayed in Fig. 1B) will abound, whereas when reduced inorganic material is abundant, photosynthesis of the type using iron—sulphur cluster acceptors (displayed in Fig. 1C) will dominate (Madigan et al., 2014).

For basic scientific research, all these three forms of 'nonoxygenic' photosynthesis (ie, those forms that do not use water as an electron donor and hence do not evolve oxygen) are important, particularly because they provide well-accessible model systems that allow detailed studies of the mechanisms operating in the more complex process of oxygenic photosynthesis. In this respect the parallels between quinone-based and iron—sulphur cluster-based reaction centres (operating in purple-sulphur and purple-non-sulphur bacteria, respectively; see Fig. 1B and C) and PSII and PSI, respectively, are very significant (two types). Regarding their biology, important unanswered basic questions still remain, particularly regarding the mutual competitiveness and coexistence of retinal- and (bacterio)chlorophyll-based photosynthesis (Hellingwerf, Crielaard, & Westerhoff, 1993).

3. 'DIRECT CONVERSION'

A closing of the global carbon cycle as referred to in Section 1 inevitably implies an intricate involvement of, and major contribution by, natural (oxygenic) photosynthesis. This process by itself is responsible for the uptake of more than 100 gigatonnes of CO_2 from the atmosphere annually (Falkowski et al., 2000), be it that a significant part of it is directly emitted back into the atmosphere via various mechanisms of photorespiration. Nevertheless, a large fraction of the carbon (and oxygen) of this CO_2 is converted into the multitude of molecules that jointly form biomass. Although only part of this biomass can be made available for human use, this is a very significant part. This is first and foremost because it provides us both directly and indirectly (ie, as feed for husbandry animals) with the food that we humans consume.

Nevertheless, a significant second use of this fixed carbon is that it provides us with new, renewable (eg, construction) materials and with renewable energy. In the development towards a sustainable society this second type of use has become so important that it has generated the 'food versus fuel' controversy (Marris, 2006) on the topic of how to best use the products of natural photosynthesis. The reason behind this is that some of the very nutritious components of plants for human consumption, such as starch and triglycerides, are also excellent starting substrates for conversion into liquid biofuel, the form of fossil energy that is most difficult to substitute with a renewable alternative. Yet society will continue to need this form of energy for aviation and heavy transport. But in addition to food and fuel, there is also a need to destine a significant part of the available biomass to the production of renewable (construction) materials.

Through the use of an approach that has been labelled by some as 'bio-raffinage' (Elbersen et al., 2003) it may be possible to bring some relief in the competition for biomass between the production processes of food, materials and fuel. This is because many forms of biomass contain parts that are less nutritional, but that are nevertheless chemically reasonably well defined and processable, such as the cellulose, the lingocellulose and the lignin fraction. A biorefinery approach can therefore be applied both to the primary product of photosynthesis, as well as to the agricultural waste that remains after food- and feed production. This approach also may be an important aspect of the strategy to make new approaches of production of renewable materials and liquid fuels economically competitive (Wijffels, Kruse, & Hellingwerf, 2013). Nevertheless, in spite of all possibilities to modify and adjust the approach of traditional crop-based photosynthesis to the future needs for production of food, materials and liquid energy carriers, the inherent maximal efficiency of plant photosynthesis — which is 4% and 6.5% for photosynthesis in C(3) and C(4) plants, respectively (Zhu, Long, & Ort, 2008), but in practice is often below 1% (Rabbinge, 1993), or even lower — will not be enough to produce sufficient supplies.

It is therefore generally assumed that further improvement in the efficiency of the application of the process of natural photosynthesis will be necessary to make a successful transition to a closed global carbon cycle. Towards this end one can follow, for example, the series of ideas discussed in the landmark paper of Blankenship et al. (2011) in which various proposals are made to modify (genetically) the basic process of oxygenic photosynthesis to allow it to use a larger part of the spectrum of electromagnetic radiation from the sun (ie, use of radiation from outside the PAR window), and also more efficiently. The latter may be achieved not only through the use of far-red absorbing chlorophylls (Blankenship et al., 2011), but possibly also via the use of red-sifted retinal-based proton pumps (Ganapathy et al., 2015). However, to be able to exploit these possibilities to their fullest extent, it will require the efforts of many, over a period of many years. Also, some of the proposed adaptations, like the adjustment of the energy gap to be bridged in PSI, may fire back on the efficiency of proton transport by the components of the central part of the Z-scheme, because these proton-pumping reactions are generally assumed to require a significant amount of excess free energy to achieve complete coupling between electron and proton transport (Finazzi, 2002). And impairment of proton pumping will directly affect the overall efficiency of photosynthesis.

For these and other reasons, we have initiated a new approach in the application of oxygenic photosynthesis for the production of materials, including liquid fuel, which has subsequently been baptized 'direct conversion' (Angermayr, Hellingwerf, Lindblad, & de Mattos, 2009). This approach implies the use of cyanobacteria, the only prokaryotic representatives that are able to carry out oxygenic photosynthesis, in engineered form to directly synthesize the preferred product. This has a number of key advantages, as summarized in Table 1. First, regarding the organisms selected: Cyanobacteria

Table 1 Summary of the Advantages of 'Direct Conversion' With Cyanobacteria Over Competing Applications of Oxygenic Photosynthesis

- Higher absolute efficiency of photosynthesis (up to 10%)
- Smaller land use than crop-based photosynthesis because of complete growth season, completely surface-covering 'canopy', higher maximal efficiency and so on.
- Much lower water use than crop-based photosynthesis.
- Reuse of minerals is straightforward through processing of biomass via methane fermentation or hydrothermal treatment of biomass.
- No involvement of inefficient plant parts such as roots and stems.
- Straightforward genetic and metabolic engineering because of prokaryotic nature of the organisms.
- Direct conversion of CO_2 into product without the need for bioraffinage.

have the highest efficiency of oxygenic photosynthesis known so far (up to 10%; see (Janssen, Tramper, Mur, & Wijffels, 2003); presumably in part because of their low maintenance energy requirement), and allow the use of simple and direct methods of genetic engineering, and relatively simple physiological engineering (eg, because of the absence of subcellular compartments). The simple and straightforward genetic engineering is crucial because 'direct conversion' implies the application of genetically engineered cyanobacteria, such that the majority of the fixed carbon is directly channelled into a preferred product like ethanol (Deng & Coleman, 1999), sucrose (Ducat, Avelar-Rivas, Way, & Silver, 2012), butyraldehyde (Atsumi, Higashide, & Liao, 2009) or lactic acid (Angermayr, Paszota, & Hellingwerf, 2012). It can be achieved by introduction of a limited number of exogenous genes that jointly form a fermentative pathway to the desired product; usually not more than two to three genes. Its main advantage is that in this approach one bypasses the inefficiencies inherent in the anabolic reactions of biomass formation and the maintenance energy required for that (Pirt, 1972), as well as the energy needed to process the resulting biomass to (a) desired product(s). Also, no energy needs to be invested into the synthesis of multicellular structures that do not contribute to the primary photosynthetic process (roots, trunks, etc.).

During the past five to six years this synthetic biology/genetic engineering approach for direct conversion of CO_2 in cyanobacteria has been widely embraced in academia. For the synthesis of a wide range of compounds, biosynthetic pathways have been added to the endogenous intermediary metabolism of particularly *Synechocystis* and *Synechococcus*, often in combination with adjustment of the endogenous metabolism to optimize the overall efficiency of the process (Atsumi et al., 2009). Meanwhile for all the four products previously mentioned, it has been demonstrated that the engineering was so successful that indeed more than 50% of the fixed CO_2 molecules is directly converted into product. This makes such engineered cells qualify as a 'cell factory' for the respective product (Fig. 2), a cell factory which carries out a 'one-pot green synthesis' (Angermayr & Hellingwerf, 2013). Indeed for some products, such as ethanol and butanediol (Duhring et al., 2013; Oliver & Atsumi, 2015) the partitioning of carbon over product and biomass has reached values of >80%. Such high carbon partitioning values significantly minimize the amount of waste biomass that is formed in parallel to the desired product and in the optimal case is processed either via anaerobic digestion or hydrothermal treatment for mineral recycling.

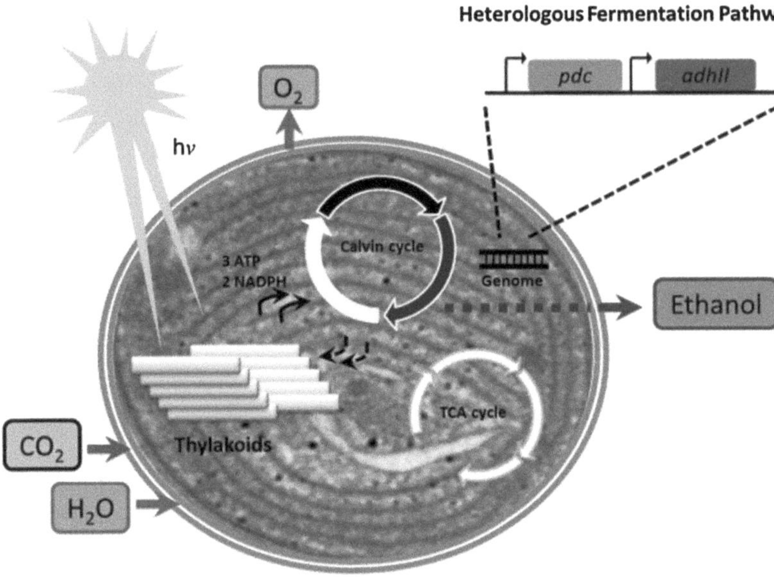

Figure 2 ***A schematic representation of a cyanobacterial cell factory.*** Exogenous genes can be added to, and deleted from, the genome of a cyanobacterium in order to achieve and/or improve the production of specific chemicals. This has led to the introduction of the term 'cell factory' where the bacterial metabolism can be transfigured to a factory assembly line. New assembly lines can also be added, already existing ones deleted or modified, in order to arrive at a more efficient production (system).

The latter approach has the advantage that the minerals become available in their oxidized form (eg, nitrate rather than ammonia), which is the most suitable form for growth of cyanobacteria.

Concentrating on literature from 2014 and 2015 one can conclude (Angermayr, Gorchs Rovira, & Hellingwerf, 2015; Oliver & Atsumi, 2014; Sarkar & Shimizu, 2015) that the range of products that can be produced via 'direct conversion' with cyanobacteria is not limited to energy- or materials-related products. Rather, the range of products that one can make is as high as with any of the traditional organisms frequently used in biotechnology. This development has been stimulated by the developments in fossil fuel prices (see the following sections) and has among others led to proof that compounds with high added value such as polyol sweeteners and flavour compounds (terpenes, alkaloids, etc.) are also well-suited products for this approach, be it that carbon partitioning to these latter products has so far been modest at most (Jacobsen & Frigaard, 2014; Kiyota, Okuda, Ito, Hirai, & Ikeuchi, 2014; Lindberg, Park, & Melis, 2010). The versatility of cyanobacteria with respect to the provision of reducing equivalents (derived directly from the electron-transfer systems in the thylakoid membrane in the form of NADPH) makes these latter organisms particularly suited for the synthesis of relatively reduced compounds such as the polyols and terpenes, because these phototrophic organisms are not subject to a closed redox balance like fermentative bacteria are, nor is the NAD(P)H subject to oxidation in a respiratory chain as in aerobic bacteria and yeasts.

The concept of 'direct conversion' can be extended to those applications in which a cyanobacterium is used for the light-driven conversion of CO_2 into some simple commodity chemical, such as sucrose, glycerol, lactic acid or glycolic acid, which then can be used by a second organism like *Escherichia coli*, to synthesize a high value—added product, for which a lot of dedicated biotechnological engineering is required that so far may not have been possible in a cyanobacterium. The two organisms can operate in separate compartments, for example, to offer an anaerobic environment to delicate (eg, oxygen-sensitive) biosynthetic enzymes (Günther et al., 2012). A mixed approach forms methane directly from algal biomass (Klassen, Blifernez-Klassen, Hoekzema, Mussgnug, & Kruse, 2015).

4. OPTIMIZATION OF 'DIRECT CONVERSION' THROUGH NATURAL PHOTOSYNTHESIS

Aspects that can be addressed with respect to optimization of 'direct conversion' through natural photosynthesis are manifold, varying from very specific strategies for one particular product, to very generic aspects of the efficiency of metabolism and growth of the cyanobacterial host. In this section we address some aspects of both. First, after a preferred product has been selected, the optimal tapping point of cyanobacterial intermediary metabolism has to be selected. Very often this will be pyruvate or fructose-6-phosphate (Angermayr et al., 2015). The next step is optimizing the enzymology and molecular biology of the heterologous pathway that must be introduced into the cyanobacterium to allow product formation (and if necessary, product export (Ducat et al., 2012)). This step includes a sufficient (but not too high, to prevent protein-burden effects) gene-expression level and enzymology of the introduced enzymes (ie, substrate affinities and molecular turnover numbers). The latter may, for example, be a problem with the more high-value added products like terpenes (Lindberg et al., 2010). Also it is important to have sufficient excess free energy dissipation in the product-forming pathway to drive product formation to near completion (Lan & Liao, 2012), for which it often helps to engineer an ATP-consuming or

CO_2-liberating step. These are the first-order priorities. These steps in most cases require detailed biochemical and metabolic engineering (eg, Angermayr & Hellingwerf, 2013; Angermayr et al., 2012; Angermayr et al., 2014). Beyond that, one can — if necessary — embark on further optimization via approaches like speeding-up metabolism via substrate channelling (eg, via the use of fusion- or cascade proteins) or via the creation of microcompartments (eg, like those involved in ethanolamine catabolism (Bobik, 2006)) if reactions need to be catalysed that are not directly compatible with the intracellular milieu of cyanobacteria.

Beyond the introduced, heterologous metabolic pathway, the organisms' intrinsic metabolism can also be optimized and optimally adjusted to the product forming pathway, to maximize the efficiency of the overall process. Cyanobacteria, in spite of their very strong specialization towards photoautotrophic metabolism, still have many redundant parallel metabolic pathways at their disposal, for example, to catalyse multiple pathways for cyclic electron transfer (Branco Dos Santos, Du, & Hellingwerf, 2014) or sugar catabolism (eg, glycolysis and the pentose-phosphate pathway). As not all these pathways have equal energetic efficiency, elimination of the least efficient ones should allow an efficiency increase. Similar arguments hold for the light-harvesting antennae of these organisms (Croce & van Amerongen, 2014). Furthermore, through these adjustments one may optimally tune the ratio of synthesis of NADPH and ATP (see the previous sections), to the requirements of the specific product formed. Accordingly, it has been argued by Knoop and Steuer (2015) that elimination of the pathways for cyclic electron transfer will increase the efficiency of formation of products like ethanol and ethylene, which only require input of NADPH as the high-energy intermediate in their synthesis and no ATP, because this elimination will decrease the ratio of ATP over NADPH generated by the cells and therefore 'force' the cells to channel more reducing equivalents to product formation (instead of making more cells).

Also selected (eg, strongly rate limiting/controlling) steps can be addressed, to remove bottlenecks in metabolism. This approach is best based on a sensitivity analysis of the main enzymes involved in product formation and can be achieved via overexpression of the relevant enzyme(s) (for published examples, see RubisCO and pyruvate kinase (Angermayr et al., 2014; Atsumi et al., 2009; Oliver & Atsumi, 2015)). Ultimately one would like to arrive at a situation in which the control over product formation is evenly distributed over all reactions involved, but this would almost require a complete drawing-board design of such a cell factory (see the following sections).

5. THE USE OF LARGE, CLOSED, OUTDOOR PHOTOBIOREACTORS

These engineered cyanobacterial cell factories will then have to be grown to produce product in large-scale photobioreactors under ambient environmental conditions of light intensity, temperature and circadian regime. Also, they will have to be incorporated into an integrated system for CO_2 supply, nutrient addition/recycling, gas (ie, O_2 and CO_2) exchange, mixing and downstream processing for product recovery, to fully exploit the remainder of the advantages listed in Table 1. Nevertheless, with respect to the actual form of the photobioreactor, a large variety of choices can (and will have to) be made. First of all, because of the use of engineered, product-forming organisms it will be unavoidable to use closed photobioreactors, rather than open ponds. The closed photobioreactors can be built in a multitude of forms, of which the basic types are: the column reactors, the flat-panel reactors and the tubular reactors (Fig. 3; for review see (Gupta, Lee, & Choi, 2015; Singh & Sharma, 2012)). Furthermore, various modes of mixing and introduction of CO_2, and stripping of excess oxygen, can be used (eg, through gas supply in so-called 'up-flow reactors' or the inclusion of a separate compartment in the reactor for gas exchange (Bentley & Melis, 2012; Molina, Fernández, Acién, & Chisti, 2001)). Nevertheless, designs will always have to embody the restriction that the path-length of sunlight through the reactor should not be more than several centimetres, because larger path-lengths will lead to complete darkness in a significant fraction of the photobioreactor. And in this nonilluminated volume only

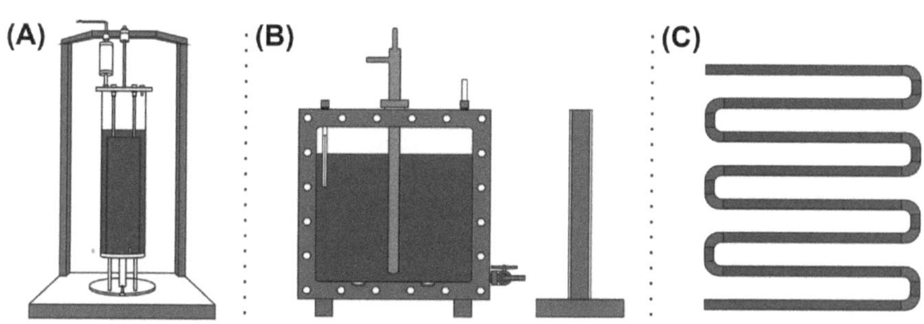

Figure 3 ***Schematic outline of three basic types of closed photobioreactor, suitable for upscaling.*** The design of three representative types of closed photobioreactors commonly employed for cultivation of cyanobacteria or algae: (A) column reactor, (B) front- and side-view of a flat-panel reactor and (C) tubular reactor.

energy dissipation can take place because of the maintenance energy requirement of the phototrophic organism.

Besides sun(light), it is also imaginable that artificial light sources, such as LEDs will be used for product formation via 'direct conversion', fuelled by renewable electricity (Chen, Yeh, Aisyah, Lee, & Chang, 2011). A speculative and simplifying calculation, which assumes that best laboratory performance now will be achievable at large-scale within a few years, leads to a very interesting conclusion: One can cover a field of nonarable land with PV panels (power efficiency 45%), and convert the electricity generated with LEDs (power efficiency 70%) to 680 nm light, which then can be used by an engineered cyanobacterium for 'direct conversion' (with 25% efficiency) in any type of photobioreactor. This then will allow photosynthesis with an efficiency higher than achievable with any current crop (ie, 8% versus 6.5% (Zhu et al., 2008)). The use of LEDs may also significantly simplify photobioreactor design (Chen et al., 2011), which may considerably increase the robustness of the overall production system (eg, regarding the possibility to create axenic conditions; see the following paragraph).

Regarding the use of the closed, large-scale, outdoor, photobioreactors, further optimization aspects are contained in the procedures to select a very robust (but accessible for molecular genetic engineering) host strain. Also, it will be of extreme importance to develop protocols for the axenic growth of these host strains in very large photobioreactors. This robustness primarily will have to be against environmental stresses, such as temperature (changes), light intensity, salinity and pH, which then in turn may greatly facilitate procedures for their large-scale axenic growth. However, these selection criteria may also have to include increased robustness against spontaneous mutation, because genetic instability of product-forming strains has occasionally been reported (Angermayr et al., 2012; Jones, 2014), particularly when plasmid-based genetic engineering is used (Borirak et al., 2015). For this, the natural recombination systems of the host organism may be targeted and/or its CRISPR/Cas systems (Kopfmann & Hess, 2013).

In 'direct conversion' one makes use of the characteristic of cyanobacteria that many low molecular weight products, such as ethanol, butanol, short-chain alkanes and monoterpenes (Angermayr et al., 2015), are rapidly secreted into the extracellular medium, and if not that this can be facilitated by engineering of a selective transporter (eg, (Ducat et al., 2012).). The product therefore can be recovered from the spent medium of the photobioreactor rather than from the biomass. This product can either be soluble, nonsoluble or volatile in the photobioreactor. In the latter two cases this will allow exploitation of the phase separation of the product, which will have a dramatic effect on the ease of its downstream processing (from either the aqueous or the gas process stream) and the associated costs thereof.

6. OUTLOOK

The key in the further development and optimization of 'direct conversion' via natural photosynthesis is a reliable computational simulation of the process. The developments in systems biology are bringing many innovative developments in the computational simulation of intermediary metabolism and growth of all forms of life at the cellular level. Prominent among these is flux-balance analysis, a constraint-based form of modelling of the stoichiometric metabolic network of a particular organism, any of several optimization criteria (Knoop & Steuer, 2015). Kinetic aspects of such simulations, however, so far have been very much dependent on these optimization assumptions. On the other hand, the most advanced forms of kinetic modelling of oxygenic photosynthesis in microorganisms use the assumption of functional PSII/PSI units (Garcia-Camacho, Sanchez-Miron, Molina-Grima, Camacho-Rubio, & Merchuck, 2012), which clearly is a crude approximation of the noninteger, often very high PSI/PSII ratio observed in many cyanobacteria. We therefore think that the way forward in these computational simulations will be to try and derive experimental values for the light-generated fluxes of NADPH and ATP and use these as input constraints in stoichiometric metabolic network models. To derive the magnitude of these fluxes from modelling electron transfer in the thylakoid membranes will be extremely complicated in wild-type cyanobacteria because a.o. of the many forms guide energy dissipation and cyclic electron transfer (Branco Dos Santos et al., 2014).

The surge in generation and harvesting of electricity has created yet another priority in society — the transient storage of electrical energy in more stable form, because of the limited capacity of batteries to do so. Also for this problem 'direct conversion' may offer a solution because of the ability of cyanobacteria to efficiently photosynthesize with LED-generated 690-nm light (which may be as high as 25—30% power conversion). This would call for a new type of volumetric (or 3D) photobioreactor in which electricity generates LED light so that everywhere in the reactor the light intensity will be in between the compensation point (Fogg, Stewart, Fay, & Walsby, 1973) and the intensity at which significant photoinhibition starts. With proper design, this would allow for the most efficient use of space and convenient operation of large-scale axenic cultures of cyanobacteria, which would be able to convert peak-shaved electricity into a liquid biofuel. In such applications living organisms like engineered cyanobacteria may turn out to be much better suited than, for example, chemical electrolysers, because they function as almost perfect rectifiers (Angermayr et al., 2012), in contrast to the chemical devices which suffer significantly from corrosion under such conditions of fluctuating loads. One could even go one step further and predict that with current laboratory-type photovoltaic cells available in the field, the most efficient form of agriculture would be to plant these photovoltaic cells on nonarable land and grow cyanobacteria or algae in such a 3D reactor. The actual efficiency of conversion of harvested sunlight into a biological product might for this type of photosynthesis turn out to be appreciably higher than for any conventional crop.

All these different aspects of 'direct conversion' make it more and more relevant to put serious effort in the design of an organism optimally equipped to the task of converting CO_2 and solar energy into a specific product. This design can be approached from multiple angles, for example, using genome reduction (to eliminate redundancy) or a bottom-up design with synthetic biology. Also natural selection for further optimization may have a role in this process. Currently, it cannot be decided yet which of these approaches will bring the best chances for success. Nevertheless, it is worth pointing out that this approach may have a very important secondary benefit — the more an organism is designed to specifically carry out one particular task under one specific environmental condition, the smaller the chances are that that organism will survive in the natural environment. Hence, it may well be that serious work to design an organism for optimally efficient 'direct conversion' also contributes to increased environmental safety of this approach.

The concept of 'direct conversion' has meanwhile been adapted by several academic and commercial researchers, leading to a situation in which the first results of commercial production of ethanol via this approach are already appearing (http://www.algenol.com/ and http://www.jouleunlimited.com/). Nevertheless, to use 'direct conversion' to compete successfully on an economic basis in the production of liquid fuel derived from fossil sources is still a major challenge. Actually many hold that this will not be possible without a significant tax on the release of fossil CO_2, because energy products are among the very cheapest products traded in our society. For that reason a large part of the research in the field of direct conversion has shifted to the production of materials with (considerably) higher added value. This is facilitated by the developments in ongoing research in cyanobacteria, which have shown that these organisms can rival with the classical workhorses of biotechnology, that is, *E. coli* and *Saccharomyces cerevisiae*, in the versatility of their ability to make a suite of products (Angermayr et al., 2015; Branco Dos Santos et al., 2014; Oliver & Atsumi, 2014). The main difference being that the two classical workhorses make their products from sugar, whereas cyanobacteria can do that from CO_2. The limitation of *E. coli* and *S. cerevisiae* that they need a closed redox balance under fermentative conditions does not apply to the cyanobacteria and hence the latter organisms are particularly suited to make the more chemically reduced products, such as fuels, sweeteners and terpene-based scents.

ACKNOWLEDGEMENTS

Q.C. is supported by a scholarship from the Chinese Scholarship Council and D.M. and K.J.H. were supported by funding from EC within the FP7 DEMA project, grant agreement no. 309086.

REFERENCES

Akram, N., Palovaara, J., Forsberg, J., Lindh, M. V., Milton, D. L., Luo, H. ... Pinhassi, J. (2013). Regulation of proteorhodopsin gene expression by nutrient limitation in the marine bacterium *Vibrio* sp. AND4. *Environmental Microbiology, 15*(5), 1400–1415.

Angermayr, S. A., Gorchs Rovira, A., & Hellingwerf, K. J. (2015). Metabolic engineering of cyanobacteria for the synthesis of commodity products. *Trends in Biotechnology, 33*(6), 352–361. pii:S0167-7799(15)00071-2.

Angermayr, S. A., & Hellingwerf, K. J. (2013). On the use of metabolic control analysis in the optimization of cyanobacterial biosolar cell factories. *The Journal of Physical Chemistry B, 117*(38), 11169–11175. http://dx.doi.org/10.1021/jp4013152.

Angermayr, S. A., Hellingwerf, K. J., Lindblad, P., & de Mattos, M. J. (2009). Energy biotechnology with cyanobacteria. *Current Opinion in Biotechnology, 20*(3), 257–263. http://dx.doi.org/10.1016/j.copbio.2009.05.011.

Angermayr, S. A., Paszota, M., & Hellingwerf, K. J. (2012). Engineering a cyanobacterial cell factory for production of lactic acid. *Applied and Environmental Microbiology, 78*(19), 7098–7106. http://dx.doi.org/10.1128/AEM.01587-12.

Angermayr, S. A., van der Woude, A. D., Correddu, D., Vreugdenhil, A., Verrone, V., & Hellingwerf, K. J. (2014). Exploring metabolic engineering design principles for the photosynthetic production of lactic acid by *Synechocystis* sp. PCC6803. *Biotechnology for Biofuels,* 7. http://dx.doi.org/10.1186/1754-6834-7-99, 99–6834-7-99. eCollection 2014.

Atsumi, S., Higashide, W., & Liao, J. C. (2009). Direct photosynthetic recycling of carbon dioxide to isobutyraldehyde. *Nature Biotechnology, 27*(12), 1177–1180. http://dx.doi.org/10.1038/nbt.1586.

Bentley, F. K., & Melis, A. (2012). Diffusion-based process for carbon dioxide uptake and isoprene emission in gaseous/aqueous two-phase photobioreactors by photosynthetic microorganisms. *Biotechnology and Bioengineering, 109*(1), 100–109.

Berg, M. J., John, L. T., & Lubert, S. (2012). *Biochemistry* (7th ed.). New York: W.H. Freeman.

Blankenship, R. E., Tiede, D. M., Barber, J., Brudvig, G. W., Fleming, G., Ghirardi, M. … Sayre, R. T. (2011). Comparing photosynthetic and photovoltaic efficiencies and recognizing the potential for improvement. *Science (New York, N.Y.), 332*(6031), 805–809. http://dx.doi.org/10.1126/science.1200165.

Bobik, T. A. (2006). Polyhedral organelles compartmenting bacterial metabolic processes. *Applied Microbiology and Biotechnology, 70*(5), 517–525.

Borirak, O., Rolfe, M. D., de Koning, L. J., Hoefsloot, H. C., Bekker, M., Dekker, H. L. … Hellingwerf, K. J. (2015). Time-series analysis of the transcriptome and proteome of *E. coli* upon glucose repression. *Biochimica et Biophysica Acta.* pii:S1570-9639(15)00165-X.

Branco Dos Santos, F., Du, W., & Hellingwerf, K. J. (2014). Synechocystis: not just a plug-bug for CO_2, but a green *E. coli*. *Frontiers in Bioengineering and Biotechnology, 2*, 36. http://dx.doi.org/10.3389/fbioe.2014.00036.

Chen, C., Yeh, K., Aisyah, R., Lee, D., & Chang, J. (2011). Cultivation, photobioreactor design and harvesting of microalgae for biodiesel production: a critical review. *Bioresource Technology, 102*(1), 71–81.

Croce, R., & van Amerongen, H. (2014). Natural strategies for photosynthetic light harvesting. *Nature Chemical Biology, 10*(7), 492–501. http://dx.doi.org/10.1038/nchembio.1555.

Danon, A., & Stoeckenius, W. (1974). Photophosphorylation in halobacterium *Halobium*. *Proceedings of the National Academy of Sciences of the United States of America, 71*(4), 1234–1238.

Deng, M. D., & Coleman, J. R. (1999). Ethanol synthesis by genetic engineering in cyanobacteria. *Applied and Environmental Microbiology, 65*(2), 523–528.

Ducat, D. C., Avelar-Rivas, J. A., Way, J. C., & Silver, P. A. (2012). Rerouting carbon flux to enhance photosynthetic productivity. *Applied and Environmental Microbiology, 78*(8), 2660–2668. http://dx.doi.org/10.1128/AEM.07901-11.

Duhring, U., Enke, H., Kramer, D., Smith, C., Woods, R. P., Baier, K., & Oesterhelt, C. (2013). *Genetically modified photoautotrophic ethanol producing host cells, method for producing the host cells, constructs for the transformation of the host cells, method for testing a photoautotrophic strain for a desired growth property and method of producing ethanol using the host cells.* patent US 2014/0011264 A1, Algenol Biofuels Inc.

Ebrey, T. G. (2002). A new type of photoreceptor in algae. *Proceedings of the National Academy of Sciences of the United States of America, 99*(13), 8463–8464. http://dx.doi.org/10.1073/pnas.142293599.

Elbersen, H. W., Sanders, J. P. M., van Dam, J. E. G., Goetheer, E., Boerrigter, H., Ree, R. V. … Welink, J. H. (2003). Transitiepad "bioraffinage". *Transitie Naar Een Duurzame Energiehuishouding*, 17.

Falkowski, P., Scholes, R. J., Boyle, E., Canadell, J., Canfield, D., Elser, J. … Steffen, W. (2000). The global carbon cycle: a test of our knowledge of earth as a system. *Science (New York, N.Y.), 290*(5490), 291–296. pii:8898.

Finazzi, G. (2002). Redox-coupled proton pumping activity in cytochrome b6f, as evidenced by the pH dependence of electron transfer in whole cells of *Chlamydomonas reinhardtii*. *Biochemistry, 41*(23), 7475–7482. pii:bi025714w.

Fogg, G., Stewart, W., Fay, P., & Walsby, A. (1973). *The blue-green algae*. London and New York: Academic Press.

Ganapathy, S., Becheau, O., Venselaar, H., Frolich, S., van der Steen, J. B., Chen, Q. … de Grip, W. J. (2015). Modulation of spectral properties and pump activity of proteorhodopsins by retinal analogues. *The Biochemical Journal, 467*(2), 333–343. http://dx.doi.org/10.1042/BJ20141210.

Garcia-Camacho, F., Sanchez-Miron, A., Molina-Grima, E., Camacho-Rubio, F., & Merchuck, J. (2012). A mechanistic model of photosynthesis in microalgae including photoacclimation dynamics. *Journal of Theoretical Biology, 304*, 1–15.

Günther, A., Jakob, T., Goss, R., König, S., Spindler, D., Räbiger, N. ... Posten, C. (2012). Methane production from glycolate excreting algae as a new concept in the production of biofuels. *Bioresource Technology, 121*, 454–457.
Gupta, P. L., Lee, S., & Choi, H. (2015). A mini review: photobioreactors for large scale algal cultivation. *World Journal of Microbiology & Biotechnology*, 1–9.
Hellingwerf, K. J., Crielaard, W., & Westerhoff, H. V. (1993). Comparison of retinal-based and chlorophyll-based photosynthesis: a biothermokinetic description of photochemical reaction centers. *Modern Trends in Biothermokinetics*, 45–52 (Springer).
Hellingwerf, K. J., & Konings, W. N. (1985). The energy flow in bacteria: the main free energy intermediates and their regulatory role. *Advances in Microbial Physiology, 26*, 125–154.
Jacobsen, J. H., & Frigaard, N. U. (2014). Engineering of photosynthetic mannitol biosynthesis from CO_2 in a cyanobacterium. *Metabolic Engineering, 21*, 60–70. http://dx.doi.org/10.1016/j.ymben.2013.11.004.
Janssen, M., Tramper, J., Mur, L. R., & Wijffels, R. H. (2003). Enclosed outdoor photobioreactors: light regime, photosynthetic efficiency, scale-up, and future prospects. *Biotechnology and Bioengineering, 81*(2), 193–210. http://dx.doi.org/10.1002/bit.10468.
Jones, P. R. (2014). Genetic instability in cyanobacteria – an elephant in the room? *Frontiers in Bioengineering and Biotechnology, 2*, 12. http://dx.doi.org/10.3389/fbioe.2014.00012.
Kim, Y. M., Nowack, S., Olsen, M. T., Becraft, E. D., Wood, J. M., Thiel, V. ... Metz, T. O. (2015). Diel metabolomics analysis of a hot spring chlorophototrophic microbial mat leads to new hypotheses of community member metabolisms. *Frontiers in Microbiology, 6*, 209. http://dx.doi.org/10.3389/fmicb.2015.00209.
Kiyota, H., Okuda, Y., Ito, M., Hirai, M. Y., & Ikeuchi, M. (2014). Engineering of cyanobacteria for the photosynthetic production of limonene from CO_2. *Journal of Biotechnology, 185*, 1–7.
Klassen, V., Blifernez-Klassen, O., Hoekzema, Y., Mussgnug, J. H., & Kruse, O. (2015). A novel one-stage cultivation/fermentation strategy for improved biogas production with microalgal biomass. *Journal of Biotechnology, 215*.
Knoop, H., & Steuer, R. (2015). A computational analysis of stoichiometric constraints and trade-offs in cyanobacterial biofuel production. *Frontiers in Bioengineering and Biotechnology, 3*.
Kopfmann, S., & Hess, W. R. (2013). Toxin-antitoxin systems on the large defense plasmid pSYSA of *Synechocystis* sp. PCC 6803. *The Journal of Biological Chemistry, 288*(10), 7399–7409. http://dx.doi.org/10.1074/jbc.M112.434100.
Lan, E. I., & Liao, J. C. (2012). ATP drives direct photosynthetic production of 1-butanol in cyanobacteria. *Proceedings of the National Academy of Sciences of the United States of America, 109*(16), 6018–6023. http://dx.doi.org/10.1073/pnas.1200074109.
Lindberg, P., Park, S., & Melis, A. (2010). Engineering a platform for photosynthetic isoprene production in cyanobacteria, using synechocystis as the model organism. *Metabolic Engineering, 12*(1), 70–79.
Madigan, T. M., Martinko, M. J., Bender, S. K., Buckley, H. D., Stahl, A. D., & Brock, T. (2014). *Brock biology of microorganisms* (14th ed.). Boston: Pearson: Benjamin Cummings.
Marris, E. (2006). Sugar cane and ethanol: drink the best and drive the rest. *Nature, 444*(7120), 670–672. pii:444670a.
Mitchell, P. (1957). A general theory of membrane transport from studies of bacteria.
Molina, E., Fernández, J., Acién, F., & Chisti, Y. (2001). Tubular photobioreactor design for algal cultures. *Journal of Biotechnology, 92*(2), 113–131.
Nocera, D. G. (2012). The artificial leaf. *Accounts of Chemical Research, 45*(5), 767–776. http://dx.doi.org/10.1021/ar2003013.
Oesterhelt, D., & Stoeckenius, W. (1971). Rhodopsin-like protein from the purple membrane of halobacterium *Halobium*. *Nature, 233*(39), 149–152.

Oliver, J. W., & Atsumi, S. (2014). Metabolic design for cyanobacterial chemical synthesis. *Photosynthesis Research, 120*(3), 249–261. http://dx.doi.org/10.1007/s11120-014-9997-4.
Oliver, J. W., & Atsumi, S. (2015). A carbon sink pathway increases carbon productivity in cyanobacteria. *Metabolic Engineering, 29*, 106–112. http://dx.doi.org/10.1016/j.ymben.2015.03.006.
Pirt, S. (1972). Introductory lecture. prospects and problems in continuous flow culture of micro-organisms. *Journal of Applied Chemistry and Biotechnology, 22*(1), 55–64.
Rabbinge, R. (1993). The ecological background of food production. In *Ciba Foundation symposium 177-crop protection and sustainable agriculture* (pp. 2–29).
Sarkar, D., & Shimizu, K. (2015). An overview on biofuel and biochemical production by photosynthetic microorganisms with understanding of the metabolism and by metabolic engineering together with efficient cultivation and downstream processing. *Bioresources and Bioprocessing, 2*(1), 1–19.
Singh, R., & Sharma, S. (2012). Development of suitable photobioreactor for algae production—a review. *Renewable and Sustainable Energy Reviews, 16*(4), 2347–2353.
Wijffels, R. H., Kruse, O., & Hellingwerf, K. J. (2013). Potential of industrial biotechnology with cyanobacteria and eukaryotic microalgae. *Current Opinion in Biotechnology, 24*(3), 405–413. http://dx.doi.org/10.1016/j.copbio.2013.04.004.
Zhu, X., Long, S. P., & Ort, D. R. (2008). What is the maximum efficiency with which photosynthesis can convert solar energy into biomass? *Current Opinion in Biotechnology, 19*(2), 153–159.

CHAPTER THREE

Bioinspired Photocatalysis

A. Aukauloo*,§,1, W. Leibl§
*Institut de Chimie Moléculaire et des Matériaux d'Orsay, UMR-CNRS 8182, Université Paris-Sud XI, Orsay, France
§Service de Bioénergétique, Biologie Structurale et Mécanismes (SB_2SM), CEA, iBiTec-S; Biochimie Biophysique et Biologie Structurale (B_3S), I_2BC, Gif-sur-Yvette, France
[1]Corresponding author: E-mail: ally.aukauloo@u-psud.fr

Contents

Abstract

Natural photosynthesis has since long been considered as the paradigm for conversion of solar energy into chemical energy. Enormous advances in the understanding of the underlying structures and mechanisms delivered blueprints for reproducing and improving the process in artificial systems. Evolution has succeeded in developing and optimizing highly efficient enzymes which perform demanding multielectron catalytic reactions with a selectivity and stereospecificity unreached by synthetic devices. In addition, the metals employed at the catalytic sites of these enzymes are abundant and cheap. Sure, the overwhelming complexity found in the biological realm appears discouraging for the engineer who engages in a biomimetic approach. However, a large part of this complexity is related to the very specific conditions and procedures of biosynthesis and regulation of these machineries in the environment of a living cell. The basic supposition of bioinspired approaches is that certain structural and functional key features recognized as crucial in the biological systems can be successfully implemented in simpler and much more robust artificial constructs, tailored to the specific requirements of a targeted technological application. Here we present representative examples of the ongoing efforts in the development of bioinspired, supramolecular complexes for production of solar fuels and discuss the main bottlenecks and problems to be solved before their application.

Advances in Botanical Research, Volume 79
ISSN 0065-2296
http://dx.doi.org/10.1016/bs.abr.2016.03.003

List of Abbreviations

DSSC	Dye-sensitized solar cell
EA	Electron acceptor
ED	Electron donor
HER	Hydrogen evolution reaction
OEC	Oxygen evolving complex of photosystem 2
PEC	Photoelectrochemical cell
PEM	Proton exchange membrane
S	Sensitizer, chromophore
WOC	Water oxidation catalyst

1. INTRODUCTION

Natural photosynthesis performs a huge energy transformation producing annually 100 TW in the form of biomass from 100,000 TW of solar energy received by our planet from the sun. In geological times, this biomass has been converted to fossil forms such as coal, oil and gas, conveniently consumable by humans. Shifting to a renewable energy supply for our societies implies production of solar fuels. This can be done by processes based on natural photosynthesis, such as transformation of biomass or engineering of photosynthetic microorganisms ((Lü, Sheahan, & Fu, 2011), see also chapter: 'Direct Conversion': Artificial Photosynthesis With Cyanobacteria by Chen, Montesarchio, & Hellingwerf, 2016). An alternative approach, called artificial photosynthesis, is the development of synthetic devices for capture of solar energy, its conversion and storage in chemical energy vectors (Fig. 1).

Whatever the transformation process might be, to be successfully employed on a significant scale, it has to fulfil certain requirements of efficiency (including selectivity), robustness, scalability and cost. As has been pointed out, the big issue is to fulfil these four requirements simultaneously (McKone, Lewis, & Gray, 2014). As an example, the indirect conversion process using photovoltaics and electrolysers to produce solar hydrogen is

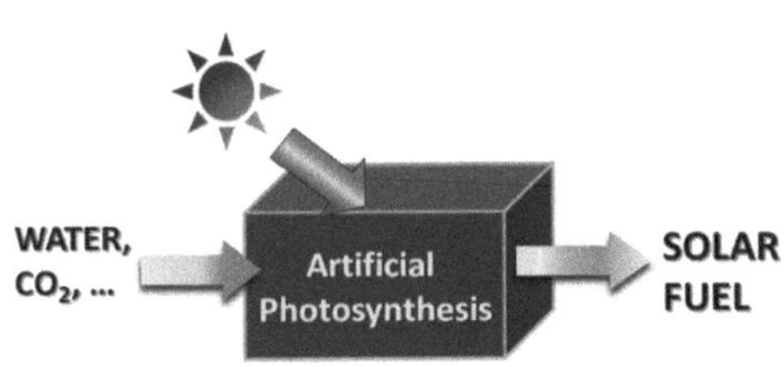

Figure 1 The 'green box' (grey in print versions) of artificial photosynthesis represents materials and processes engineered to convert abundant raw materials into energy carriers using solar light as the input of renewable energy.

reasonably efficient, robust and scalable but is actually too costly to be deployed outside some niche applications. Direct, integrated photocatalytic or photoelectrochemical conversion processes promise potentially higher efficiency and lower cost, but suffer so far from scalability and, especially, durability (Coridan et al., 2015; Jacobsson, Fjallstrom, Edoff, & Edvinsson, 2014; Rodriguez, Modestino, Psaltis, & Moser, 2014).

In photosynthetic organisms such as plants, cyanobacteria and algae, sunlight is used to drive the four-electron and four-proton oxidation of water. The electrons and protons recovered from this reaction are stored as equivalents of H_2 on carriers of reducing equivalents, such as NADPH, to ultimately reduce CO_2 to sugars. The whole process is clearly separated in three different, consecutive subprocesses (Fig. 2). *Capture* of photonic energy due to absorption by a variety of pigments giving rise to formation of excited states, *conversion* of these excited states into redox power by a charge separation process fast enough to compete with intrinsic deactivation pathways and finally *catalysis* using stored redox equivalents to form or break chemical bonds.

An important feature of the natural photosynthetic systems is that these three basic functions are performed by separate units. Chromophores, electron donor (ED) and acceptor groups, and catalytic sites are molecular-based, spatially organized modules. Bioinspired approaches to artificial photosynthesis are usually based on the same principle of modular design (see Fig. 3 for an example). In addition, the target is to incorporate another

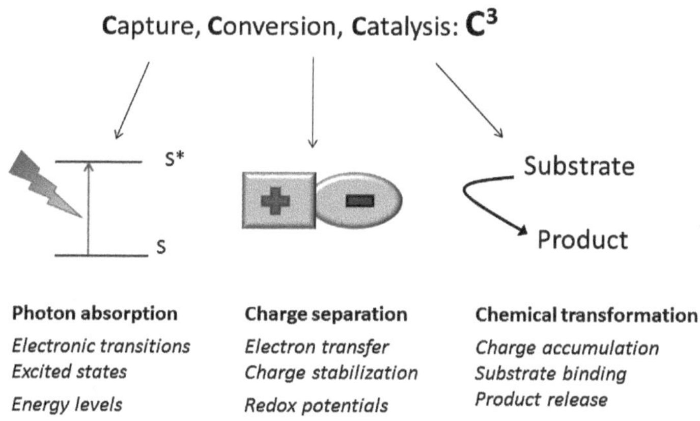

Figure 2 The three basic functions of photocatalytic systems and related processes and properties.

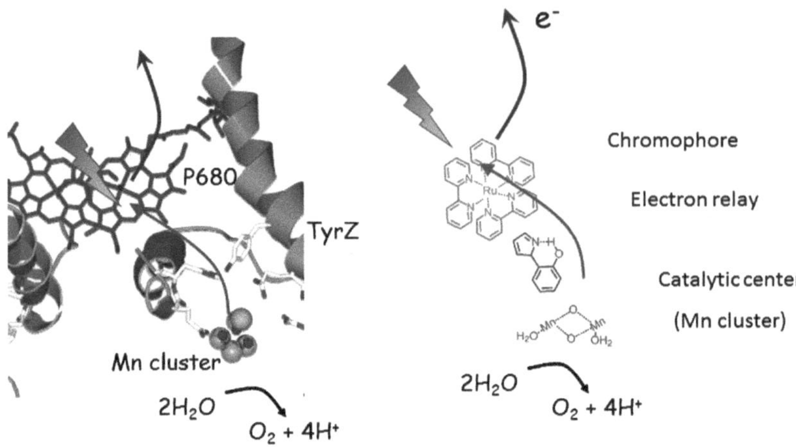

Figure 3 Donor side of the natural, light-driven water-oxidizing enzyme (photosystem 2) and biomimetic model comprised of chromophore, electron relay and Mn-based catalyst modules.

important feature of the natural systems with the use of nonnoble metals especially for the catalytic sites. While the best catalysts for water oxidation and proton reduction used, for example, in electrolysers and fuel cells are based on rare and expensive metals such as iridium, palladium and platinum, enzymes such as the oxygen-evolving complex of photosystem 2 (OEC) and hydrogenases perform the same reactions very efficiently with manganese, iron or nickel in their catalytic sites. In the former systems, it is mainly the 2-dimensional surface structure which determines substrate and product binding and catalytic activity. On the other side, it is clear that in the natural systems, catalytic activity and selectivity are largely determined by the ligands to the nonnoble metal centres as well as by the 3-dimensional architecture including the surrounding protein scaffold which controls substrate access and provides the correct electrostatic and dielectric environment as well as electron and proton transfer pathways. To establish similar performance in synthetic molecular constructs requires at least optimization of the first and second coordination sphere of the metal centres of the catalysts. As a matter of fact, bioinspired synthesis does not aim to reproduce all the enormous complexity of the biological systems which is largely imposed by the very particular conditions and restricted resources available in living organisms. The assumption is that much simpler and more stable model systems can be developed without too much penalty with respect to performance. The difficulty then is to define the level of complexity which is necessary and sufficient for the desired function.

2. STRATEGIES

The advantage of a molecular approach to photocatalyst design is the possibility of virtually unlimited rational variation of the molecules synthesized (Ashford et al., 2015; Berardi et al., 2014). Carefully chosen substitutions allow tuning redox and optical properties. Rates of intramolecular energy and electron transfer reactions can further be tuned by varying distances and electronic coupling between modules. However, synthetic chemistry, often long and tedious, has to be guided. Complexes designed in a first place with inspiration from natural models need to be characterized on a functional level to understand the events induced by light excitation and to identify unwanted side reactions, rate-limiting reaction steps and degradation pathways which represent bottlenecks for performance. Typically for this characterization, a wide range of electrochemical and spectroscopic methods is employed. In particular time-resolved studies, using synchronization with the help of short laser flashes allows to follow the fate of the excited state, determine the rates of charge transfer reactions and identify unstable intermediates. The knowledge on a mechanistic level thereby gained is the very basis of the attempt to design and synthesize modified versions in an iterative process of improvement (Fig. 4).

It is important to keep in mind that any catalytic process for fuel production encompasses both, oxidative and reductive catalysis. A fuel is a reduced

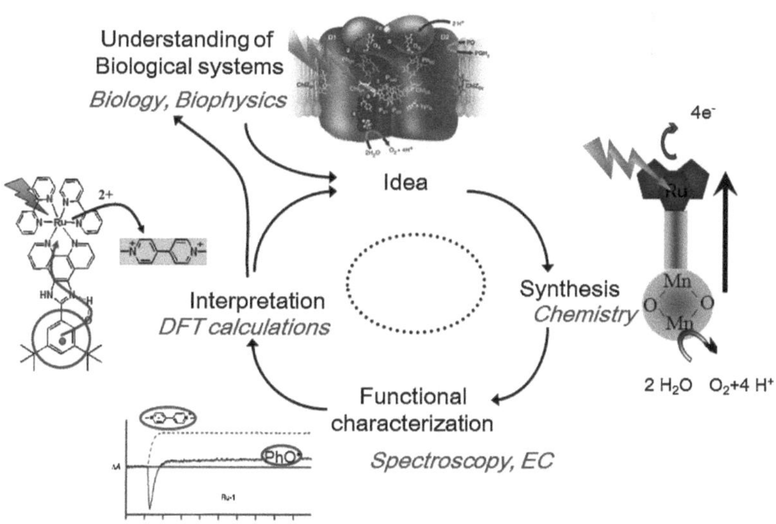

Figure 4 Iterative, bioinspired approach to photocatalyst design.

chemical substance from which the energy stored in chemical bonds can be recovered by reaction with oxygen. The process of production of a fuel implies a reductive process with storage of hydrogen equivalents on a carrier molecule. Importantly this reductive process relies on the supply of electron and protons which have to be produced by oxidation of an abundant resource: water. It appears clearly that this oxidative part, the activation of such a stable molecule as H_2O, actually represents a significantly bigger challenge than the catalysis of the reduction reactions.

The simplest design for a photocatalyst for water photolysis is shown in Fig. 5. Similar to the architecture of photosynthetic reaction centres, electron acceptor (EA) and electron donor (ED) groups are connected to opposite sides of a chromophore (S). Upon excitation of the chromophore, charge separation occurs by intramolecular electron transfer (ET) from the chromophore excited state to the EA followed by electron transfer from the ED to the oxidized chromophore regenerating the ground state of the

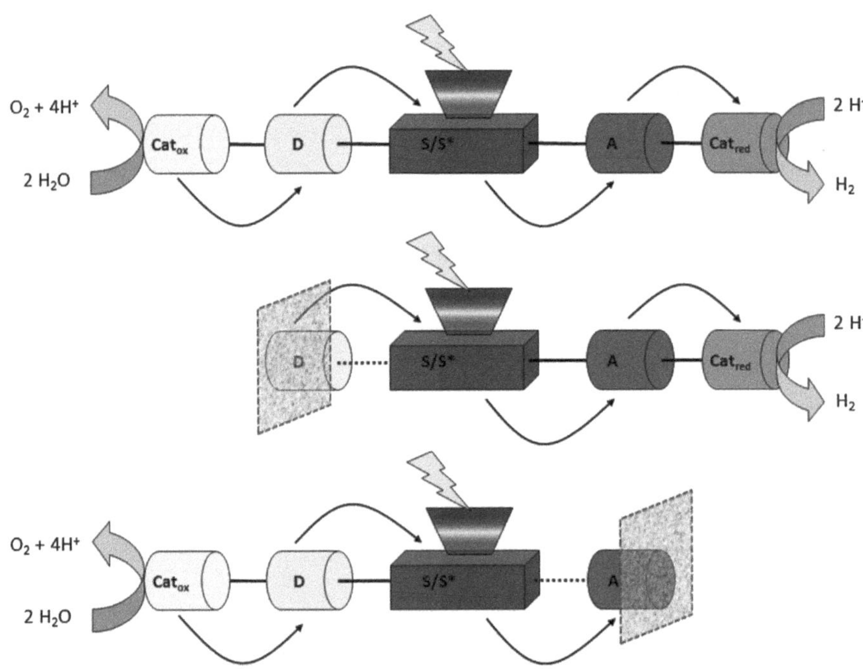

Figure 5 Top: Schematic representation of a supramolecular modular photocatalyst constructs for water photolysis. Middle: Photocatalyst for reduction with sacrificial electron donor or as part of a photocathode. Bottom: Photocatalyst for oxidation with sacrificial electron acceptor or as part of a photoanode.

latter. The reduced EA and the oxidized ED perform reduction and oxidation of the two catalytic sites in a thermal charge transfer reaction.

Although from a thermodynamic point of view it is possible to activate both a water oxidation catalyst (WOC) and a catalyst for proton or CO_2 reduction with the energy delivered by one photon in the visible range, there seems to be a general agreement that the more promising solution is to separate the two half reactions and supply each of the two catalysts with its light absorption and charge separation unit (Fig. 5). This strategy is also realized in the Z-scheme of natural oxygenic photosynthesis where photosystem 1 creates reducing equivalents at low potential and is fed with electrons from the water-oxidizing photosystem 2 via an electron transport chain. In this way, photocatalyst complexes for oxidative processes and reductive processes can be developed and optimized separately replacing the missing part by sacrificial EAs or donors, respectively. The role of ED or acceptor can be played by (semiconducting) electrodes to which the complexes are grafted via their chromophore module (Fig. 5). Each of the designed photoanodes and photocathodes can again be tested separately in a photoelectrochemical three-electrode setup. Photoanodes and photocathodes can then be combined in a photoelectrochemical cell (PEC) as the final device (Fan et al., 2014; Li et al., 2015; Sato, Arai, & Morikawa, 2015). Such devices have many properties in common with the regenerative dye-sensitized solar cells (DSSC) the development of which has known enormous progress since the pioneering work of Michael Grätzel (Fakharuddin, Jose, Brown, Fabregat-Santiago, & Bisquert, 2014; Hardin, Snaith, & McGehee, 2012; Odobel & Pellegrin, 2013; Oregan & Gratzel, 1991; Swierk &

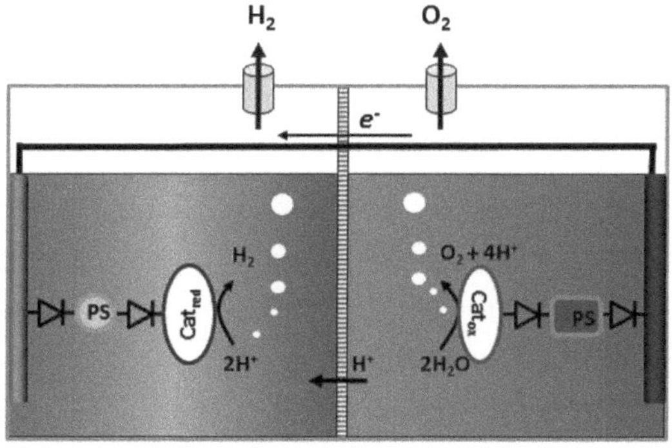

Figure 6 Photoelectrochemical cell as the target device for solar water splitting. The importance of directional electron transfer is symbolized by the diode symbols.

Mallouk, 2013). Like in fuel cells, anode and cathode are electrically connected to convey electrons from the oxidation to the reduction side while protons are transferred via a proton exchange membrane (Fig. 6). In contrast to DSSCs, no redox-active electrolyte is employed. As this component is one of the efficiency-limiting factors for DSSCs, photoelectrochemical cells could in principle permit higher energy conversion efficiencies (Fig. 6).

3. CHROMOPHORES

In a bioinspired approach to artificial photosynthesis, the function of light absorption is realized in separate chromophores. Many catalysts are coloured compounds that can combine chromophore and catalyst function in one molecule (see Funyu, Isobe, Takagi, Tryk, & Inoue, 2003 for an example). Although such systems have the advantage of simplicity, an important drawback is that their visible light absorption properties are dependent on the oxidation state of the metal, making it difficult to maintain constant visible absorption properties as the metal changes oxidation states during the catalytic cycle. Therefore the use of dedicated chromophores is preferable to achieve efficient coupling of the one-photon, one-electron events on the sensitizer to the multielectron events on the catalyst. In this case, in general, the excited state of the chromophore is quenched by ET to EAs or donors, creating the oxidized or reduced state of the chromophore which in turn activates via intramolecular ET catalysts for oxidation and reduction catalysis, respectively. EA and ED functions can be realized by external sacrificial molecules in solution, by redox-active groups covalently linked to the chromophores, or by semiconducting electrodes, to which the photocatalysts are grafted.

The choice of suitable chromophores, often also called sensitizers, for molecular photocatalysts is mistakenly considered as not an urgent problem. In fact, this research theme is the focus of several groups dedicated to the quest for cost-effective chromophores. For the development of catalysts activated by light, the most important aspect is the spectral region of photons absorbed, which should be in the visible range, rather than in the UV. A plethora of sensitizers can be found in the literature, ranging from metal-coordination complexes to organic dyes (and quantum dots). Certainly the most widely employed model chromophores are of the Ruthenium-polypyridyl family (Juris et al., 1988; Kalyanasundaram,

1982). Besides representing a convenient platform for the synthetic chemist, these complexes feature absorption in the visible range (typically up to 500 nm). Importantly, the triplet excited state shows a relatively long life time in the 0.5−1.5 μs range and is both a good oxidant and reductant making this chromophore suitable for oxidation and reduction processes. By functionalization of the ligands, absorption and redox properties can be tuned within a certain range. Similar coordination complexes based on iridium, osmium, platinum or rhenium are mainly employed to sensitize reductive reactions (Archer & Weinstein, 2012; Gärtner et al., 2011; Wenger, 2009).

To avoid the use of (noble) metals, alternative chromophores are needed. An iron-nitrogen-heterocyclic-carbene sensitizer has been reported with sufficiently long excited state life time to efficiently inject electrons into TiO_2 nanoparticles (Harlang et al., 2015). Provided that the oxidation potential can be increased to be sufficient to drive WOCs, such complexes could become an alternative to the classical Ru-based sensitizers. Copper(I) bisbipyridine-type complexes are also under investigation for their photophysical properties. Metal-free organic dyes are another group of sensitizers used to drive catalysts. As for the ruthenium-polypyridyl-type sensitizers, the development of inorganic dyes is an extremely active field in the realm of DSSC. Free-base and Zn-porphyrins are one example, mainly suited for reductive processes. They show higher extinction coefficients than metal coordination complexes and absorb over a wider range of the visible spectrum. Examples for other commonly used inorganic dyes include different variants of BODIPY dyes, Eosin and Rose Bengal (Fig. 7) (Lazarides et al., 2009; Zhang et al., 2010; Zheng et al., 2015).

Another class of sensitizers is nanostructured semiconductors such as CdS nanocrystals or CdSe quantum dots. Combined with molecular catalysts, such material has been used as sensitizers in photocatalytic hydrogen generation and the construction of photocathodes (Han, Qiu, Eisenberg, Holland, & Krauss, 2012; Ruberu, Dong, Das, & Eisenberg, 2015). Such hybrid systems, combining semiconducting or metal oxide materials with molecular systems, can be considered an interesting alternative to a purely molecular approach as it allows combining the often higher stability of materials with the better variability of molecular constructs.

In natural photosystems, a large number of antenna pigments contribute to excitation of every reaction centre permitting the photocatalytic processes to proceed close to its maximum rate. Most artificial photocatalyst systems reported in the literature contain only one chromophore unit

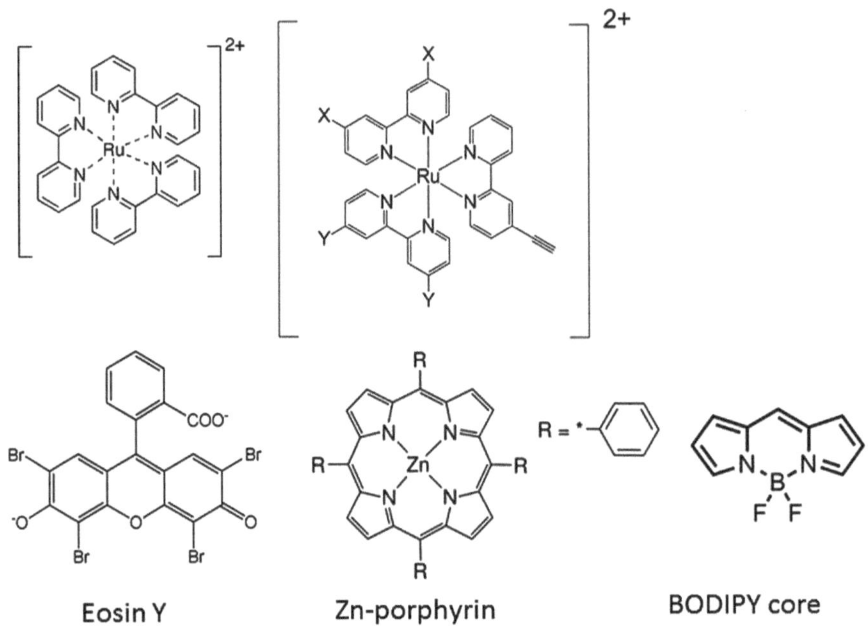

Figure 7 Top: Structure of $[Ru(bpy)_3]^{2+}$-type chromophores. Different substitutions (X and Y) allow tuning of the oxidation potential of the Ru^{III}/Ru^{II} couple by more than 400 mV. Bpy ligands can be functionalized with an alkene group for a click-chemistry reaction. Bottom: Examples of organic chromophores.

per catalyst but an increase in absorption cross section is highly desirable to reduce the time necessary for charge accumulation (and storage!) on the catalyst (A typical chromophore will absorb less than one photon per second in full sunlight; see also (Inoue et al., 2011).). Different strategies can be envisioned to increase the rate of catalyst activation. Light harvesting can be improved by panchromic absorbers realized by combination of different chromophores (Alexy, Chandrashaker, & Lindsey, 2014; Castellano, 2015). Also, additional absorbers transferring energy to a central chromophore are investigated creating artificial antenna-reaction centre systems (Kodis et al., 2006; Panda, Ladomenou, & Coutsolelos, 2012). Alternatively it has been shown that accumulation of multiple oxidative equivalents can be achieved by cross-surface ET between a catalysts and neighbouring chromophores when anchored to TiO_2 (Song et al., 2013). High chromophore to catalyst ratios might favour efficient catalyst activation.

As mentioned earlier, it is important to realize that, from a practical point of view, efficient water splitting by molecular photocatalysts is only possible if the photoactive complexes are fixed on semiconductor electrodes. Because it is the excited state of the chromophores which interacts with the semiconductor (by injecting electrons into the conduction band or extracting holes from the valence band), the chromophore part of the photocatalysts has to be functionalized with suitable anchoring groups. Carboxylate groups on the periphery of dyes are commonly used for DSSCs, but these groups show insufficient stability in aqueous solvent. Ruthenium-bipyridyl complexes with phosphonate anchoring groups appear significantly more stable (Bae et al., 2004; Hanson et al., 2012). However, there clearly is still a need for improvement of the stability in the presence of water of photocatalyst grafting to semiconductor oxides, for example by introduction of hydrophobic chains on the chromophores (Wang et al., 2004). Here again, progress could come from the field of DSSCs, where there is renewed interest in replacing organic solvents by water (Bella, Gerbaldi, Barolo, & Gratzel, 2015).

4. MOLECULAR CATALYSTS FOR OXIDATION

The four-electron oxidation of water to give dioxygen using visible light is by far the most challenging reaction in the field of artificial photosynthesis. This is even more blatant in the domain of molecular chemistry when compared to metal oxides or semiconducting materials. Not surprisingly, based on the fact that ruthenium oxide is known as one of the best catalyst used in PEM water electrolysers, molecular ruthenium-based complexes have been on the forefront of this research. The Meyer's group has pioneered this field with the discovery in 1985 of the *cis, cis*-$[(bpy)_2(H_2O)Ru(\mu\text{-}O)Ru(H_2O)\text{-}(bpy)_2]^{4+}$ complex, commonly known as the 'blue dimer' catalyst (see Fig. 8). This complex topologically holds two aqua groups which are bound to two ruthenium centres linked via a μ-oxo bridge. It was reported that this complex was capable to chemically activate the starting $H_2O\text{-}Ru^{III}\text{-}O\text{-}Ru^{III}\text{-}OH_2$ derivative to $O{=}Ru^{V}\text{-}O\text{-}Ru^{V}{=}O$ through a $4e^- + 4H^+$ oxidation process. Mechanistic insights indicate that the resulting Ru—oxo complex would undergo a so-called nucleophilic attack from a water molecule to form the intermediate $HO\text{-}Ru^{IV}\text{-}O\text{-}Ru^{IV}\text{-}OOH$, and therefrom release oxygen with a TON of around 10. An important step in molecular photocatalysis was accomplished by using light energy to power that reaction thereby excluding the use of the powerful Ce^{4+} as oxidant. Under visible light irradiation using $Ru(bpy)_3{}^{2+}$ as

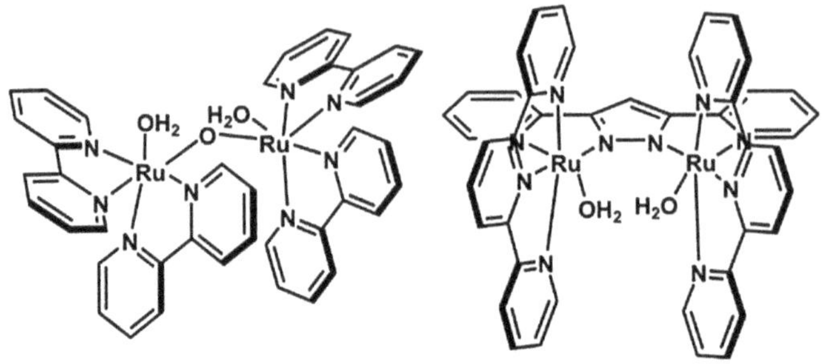

Figure 8 Dinuclear ruthenium-based catalysts. Meyer's blue dimer (left), and Llobet's *in,in*-Ru-Hbpp complex.

chromophore and $S_2O_8^{2-}$ as EA, this complex generates about 100 equivalents of oxygen during the first 15 min of the reaction. It was argued that the loss of photocatalytic activity originated from the chemical alteration of the chromophore (Rotzinger et al., 1987).

Based on these first findings, different groups have developed modified derivatives of the blue dimer system with the target to improve the catalytic activity of ruthenium-based complexes. Fruitful chemical modifications of the original blue dimer concern the replacement of the single oxo atom bridge with a binucleating pyrazole ligand (Fig. 8), as demonstrated by the *in,in*-$\{[Ru^{II}(tpy)\text{-}(H_2O)]_2(\mu\text{-bpp})\}^{3+}$ (bpp = 3,5-bis-(2-pyridyl) pyrazolate) dinuclear ruthenium complex where the aqua ligands are orientated in a face-to-face position at 2.48 Å to each other. In a similar fashion, the initial $H_2O\text{-}Ru^{II}\text{-bpp-}Ru^{II}\text{-}OH_2$ species can be chemically oxidized to $O{=}Ru^{IV}\text{-}O\text{-}Ru^{IV}{=}O + 4e^- + 4H^+$. However, an interesting finding here is that this di-oxo species evolves through an intramolecular μ-1,2-peroxo species resulting from the O—O bond formation that ultimately evolves oxygen (Bozoglian et al., 2009; Romero et al., 2008; Sala, Romero, Rodriguez, Escriche, & Llobet, 2009). These are clear examples in literature where well defined—chemical modulation in the design of a ligand can direct the mechanism of a catalytic reaction. Other systems holding negatively charged coordinating groups have been elaborated with the target to stabilize the high oxidation states of the ruthenium centre. A dinuclear ruthenium complex with a *cis* structure bearing a ligand 1,4-bis(6′-COOH-pyrid-2′-yl)phthalazine with carboxylate groups was shown to shift the oxidation potentials of the ruthenium centre to lower values as compared to neutral ligands. This complex was also shown to oxidize water in the presence of a powerful oxidant as discussed earlier. Elegant synthetic design of more sophisticated ligands has been undertaken, and their catalytic properties towards water oxidation reaction have been reported.

An interesting breakthrough in this chemistry was the finding that mononuclear ruthenium complexes could perform the four-electron and four-proton oxidation of water. Meyer and other groups have evidenced that a single-site ruthenium complexes could indeed realize this challenging task of ripping four electrons from two molecules of water in a catalytic manner. Mononuclear ruthenium complexes such as $Ru^{II}{-}H_2O$(tpy)(bpy) (Fig. 9) and $Ru^{II}{-}H_2O$(tpy)(bpm) (bpz = pipyrazine, bpm = bipyrimidine) were reported to chemically catalyse the oxidation of water with good TON and TOF (Concepcion et al., 2009; Concepcion, Jurss, et al., 2010; Concepcion, Jurss, Templeton, & Meyer, 2008; Concepcion, Tsai, Muckerman, & Meyer, 2010). The mechanism for these mononuclear complexes is similar to that proposed by Meyer for the blue dimer. Loss of two electrons and two protons from the LL′-Ru^{II}-OH_2 leads to the formation of LL′-Ru^{IV}=O, one-electron oxidation yields LL′-Ru^{V}=O which undergoes nucleophilic attack from a water molecule to form a LL′-Ru^{III}-OOH intermediate, followed by proton loss and one-electron oxidation to form LL′-Ru^{V}-OO and release of oxygen.

Here too, the synthetic tactic of introducing carboxylate coordinating functions on the ligand scaffold was successful in stabilizing the highly oxidized ruthenium species. Compared to similar compounds with neutral ligands, insertion of carboxylato groups in the ligand structure lowers the oxidation potentials of the ($Ru_2^{II,II}/Ru_2^{II,III}$) and ($Ru_2^{II,III}/Ru_2^{III,III}$) couples to 0.903 and 1.396 V versus NHE, respectively. Under oxidation by Ce^{4+}, this catalyst is capable of sustained H_2O oxidation for 20 h with a TON of 10,400 and TOF of 1.2 s^{-1}. Other anionic ligands have also been demonstrated to function in a similar fashion. A further added chemical feature to

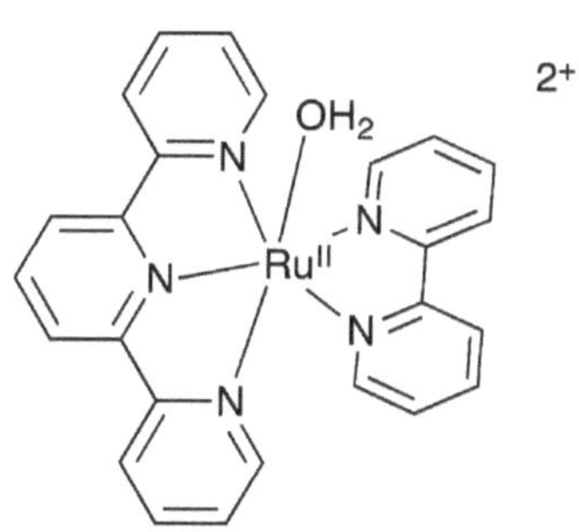

Figure 9 $[Ru^{II}\text{-}H_2O(tpy)(bpy)]^{2+}$ as single-site catalyst for water oxidation reaction.

develop hydrogen-bonding interaction with the metal-bound water molecule has also shown to enhance the catalytic activity of this family of ruthenium complexes (Matheu, Ertem, et al., 2015). The involvement of proton-coupled electron transfer processes in the water oxidation reaction supports this interesting catalytic pattern. This type of sophistication in the design of molecular catalysts is a clear example of the synergy between biology and chemistry. Indeed, a series of amino acids are known to be involved in conveying protons away from the Mn_4Ca catalytic unit and at the same time participate through hydrogen-bonding interactions to direct and orientate the water molecules at the metal centres of the catalytic core.

In general, photoactivation of these catalysts using light of wavelengths >400 nm in the presence of $Na_2S_2O_8$, and $Ru(bpy)_3^{2+}$ chromophores as the sensitizers yielded oxygen with modest TON. The short duration of the photocatalytic reaction seems to be inherently linked to the EA used. Cobalt pentammine releases ammonia, which increases the pH of the reaction medium. Decomposition of both the chromophore and catalyst molecules is observed after a few minutes. Using persulfate as sacrificial EA also leads to deactivation of the photocatalytic system due to the lowering of the pH in the resulting solution.

On the outskirts of this search towards the use of ruthenium molecular photocatalysts for the water oxidation process, a mononuclear ruthenium complex originally developed by Meyer and collaborators was used in the elaboration of a membrane-electrode assembly of a PEM electrolyser cell. The idea was to test whether such molecular complexes could replace the common ruthenium oxide as catalyst at the anodic side of the PEM electrolyser. Polarization curves obtained from such hybrid cells were compared to the cells where IrO_2 or RuO_2 are used as electrodes for the anodic process (Fig. 10). Results from this study tend to support that the molecular catalyst is indeed active at the anodic electrode (Mbemba Kiele et al., 2013). However, an important missing piece of information from this study concerns the actual nature of the catalytic material during electrolysis. A more detailed study on the functionalization of graphitic electrode with molecular mononuclear ruthenium complexes indicates the formation of RuO_2 particles at the surface of the electrode during electrocatalysis (Matheu, Francàs, et al., 2015).

An intriguing paradox that chemists face is the ongoing development of noble metal complexes as WOC, while at the same time it is constantly argued that we need to move to cheaper and more abundant material as the *sine qua none* condition to implement new discoveries on a large scale.

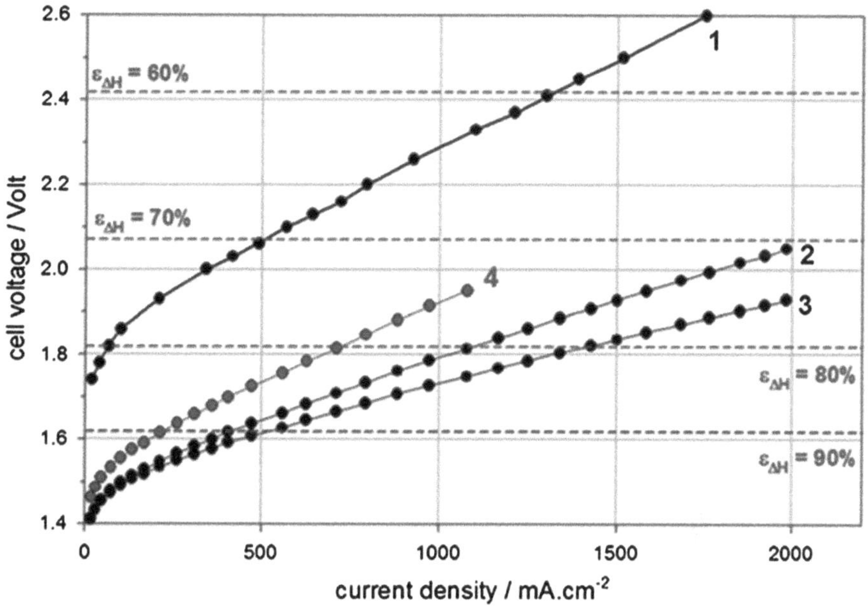

Figure 10 Polarization curves measured on (1) Pt/Nafion117/Pt; (2) Pt/Nafion117/Ir; (3) Pt/Nafion117/RuO_2; (4) Pt/Nafion117/Vulcan-[Ru(tpy)(bpy)(H_2O)]$^{2+}$. *Reprinted from Mbemba Kiele, N., Herrero, C., Ranjbari, A., Aukauloo, A., Grigoriev, S. A., Villagra, A., & Millet, P. (2013). Ruthenium-based molecular compounds for oxygen evolution in acidic media.* International Journal of Hydrogen Energy, *38*(20), 8590–8596. with permission from Elsevier.

Therefore, we make no mistake in saying that we need to devote much attention to the chemistry of first-row transition-metals although we know pertinently that first-row transition-metal catalysts are plagued by their low stability and their inherent kinetic lability in comparison to their second- and third-row transition congeners. We discuss here on the recent findings where metal complexes with first-row transition elements have been reported as WOC. Although the catalytic activities of these metal complexes are far from any technological deployment, we have the feeling that these discoveries may gather much momentum and go on to a next stage to compete with well-established noble metal catalysts.

In a recent excellent review Åkermark et al. published an exhaustive state of the art on molecular catalysts for oxidation of water (Kärkäs, Verho, Johnston, & Åkermark, 2014). Herein, our idea is to give the broad panorama of the main incursions in this quest. Perhaps, the Brudvig–Crabtree manganese di-μ-oxo complex published in the early 1990s has been the booster for molecular chemists warranting that molecular manganese complexes could oxidize water (Chen et al., 2005; Limburg, 1999; Young,

Brennan, Tagore, & Brudvig, 2015). Dismuke's tetranuclear manganese complex was another example again supporting the molecular approach with first-row transition ions (Dismukes et al., 2009). However, investigations in 2015 have shown that chemical alterations of these molecular complexes lead to the formation of oxides that act as the catalytic species (Suseno et al., 2015). Such alterations of the initial molecular catalysts leading to the departure from homogeneous to heterogeneous catalysis with the formation of nanoparticulates have cast serious interrogations on the real nature of the catalytic species in molecular-based chemistry. Importantly, before proposing any catalytic route even sometimes backed by theoretical calculations, great care should be taken to analyse the fate of the chemical precursors. Moreover, these chemical pathways must be understood in order to guide the synthetic chemists in their tasks to elaborate more robust and efficient systems. For this purpose, a series of physical methods are now being currently used to monitor photocatalytic or electrocatalytic reactions. Transmission electron microscopy (TEM) and energy dispersive X-ray spectroscopy are analytical methods that are commonly used to detect and chemically analyse the nature of nanoparticles at the surface of electrodes eventually formed upon electrocatalysis. Characterizing most finely distributed particles and their relative morphology is critical to the understanding of the catalytic performance. STEM HAADF (scanning transmission electron microscopy high-angle annular dark field) tomography is the ideal tool to characterize nanoactive sites around 1 nm or less in size. X-ray photoelectron spectroscopy also helps to determine the oxidation states of the metal ions within the electrode-deposited material. Chemical modifications of the molecular precursors in solution under catalytic conditions are followed by Dynamic Light Scattering techniques that provide with the size distribution of the nanoparticles.

Interestingly, it can be noted that synthetic models of the OEC were confined to manganese chemistry because of their presence in the OEC. Only recently the other first-row transition-metal complexes have been investigated as potential WOCs. Examples of manganese, iron, cobalt and copper complexes have been reported as molecular catalysts for the water oxidation reaction. These complexes can be classified within two families, one with anionic ligands meant to stabilize high oxidation states, and the other with neutral ligands.

T.J. Collins has developed a class of tetraanionic macrocyclic ligands (TAML) that can stabilize high-valent manganese and iron metal complexes. These complexes have successfully been used as hydrogen peroxide activators to generate strong oxidizing Fe(V)-oxo species, for instance, that were capable to degrade recalcitrant pollutants (Collins, 2002). Such an iron complex was shown to act as a WOC on the addition of Ce^{4+} as oxidant. A similar proposal of a water molecule attacking the electron-deficient oxygen atom at the highly oxidized iron(V) centre was presented for the formation of the O—O bond formation. An important advancement towards photooxidation of water was published by C. Panda et al. in 2014. In this study, the authors used the classic mixture of a $[Ru(bpy)_3]^{2+}$ chromophore, a sacrificial EA and an iron complex with a biuret derivative of the TAML ligand (Fig. 11). Following an exhaustive investigation and characterization, the authors demonstrated that they could photoproduce the active Fe(V)-oxo species stable at room temperature that allowed the oxidation of water. The formation of deleterious iron oxide particles was excluded based on concentration effects and DLS experiments. On a fundamental basis, such studies bring solid concurring evidence that a single-site first-row transition-metal complex can act as a catalytic centre for the water oxidation reaction. Furthermore, on a mechanistic basis, the formation of the O—O bond following a so-called 'nucleophilic attack' brings an additional example in literature to fuel the discussion on the ways the O—O bond is formed.

Other iron complexes with neutral ligands have also been reported to catalyse the water oxidation reaction in the presence of Ce^{4+} as oxidant. Interestingly, investigations on the photocatalytic properties of a mixture of a chromophore, EA and iron complex lead to the oxidation of organic sulphide substrates through an oxygen atom transfer (OAT) reaction. No

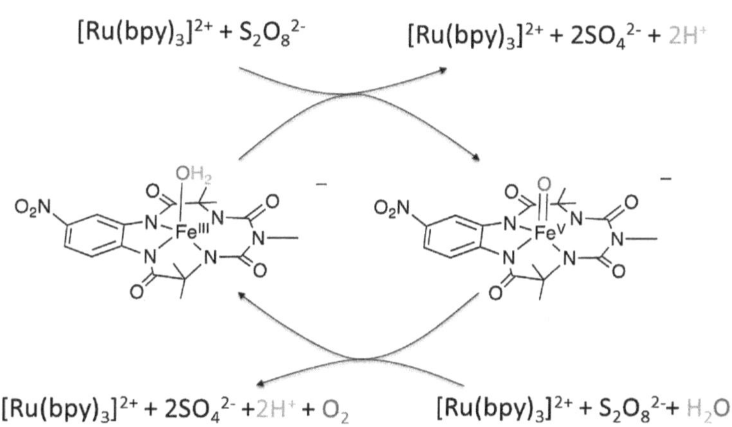

Figure 11 Photocatalytic cycle for the Biuret TAMLFe(III)Cl complex in the presence of $[Ru(bpy)_3]^{2+}$ and $S_2O_8^{2-}$.

photooxidation of water has yet been reported using this family of iron complexes with neutral ligands. Another study points to the formation of Fe_2O_3 particles upon photocatalytic runs with different iron complexes (Chen, Chen, Ng, Man, & Lau, 2013).

Another line of research of bioinspired photosynthesis is to use water as the oxygen atom donor to perform photooxidation of organic substrates. This reaction can be looked at as the half way activation of a water molecule generating an oxidizing M-oxo species that can ultimately transfer the oxygen atom to an organic substrate. Doing so, two electrons and two protons are theoretically recovered from the activation of a metal-bound water molecule. Different systems have been reported with such a target. A covalently linked Ru-Fe assembly holding a $[Ru(bpy)_3]^{2+}$ chromophore and a nonhaemic iron complex surrounded by a neutral ligand was reported (Fig. 12). In this study, the authors demonstrated the formation of a Fe(IV)-oxo species starting from the Fe(II)-H_2O precursor under irradiation and in the presence of an electron scavenger (Herrero et al., 2015). The generated Fe(IV)-oxo was capable of transferring an oxygen atom to a substrate. Other authors have designed supramolecular assemblies with a copper catalytic unit (Iali et al., 2015). Photocatalytic studies indicate oxygen atom transfer.

Cobalt has played a pivotal role in the research for WOCs. Nocera's group has made an enormous contribution in this field with the discovery of cobalt oxides nanoparticles as efficient catalysts for the WO reaction (Nocera, 2012). Molecular complexes with cobalt centre have been reported as capable to drive the electrochemical and chemical oxidation of water. Here too, interrogations remain on the actual integrity of the molecular precursor during the catalysis (Stracke & Finke, 2011; Wasylenko, Palmer, Schott, & Berlinguette, 2012). In a recent article, a dinuclear cobalt complex

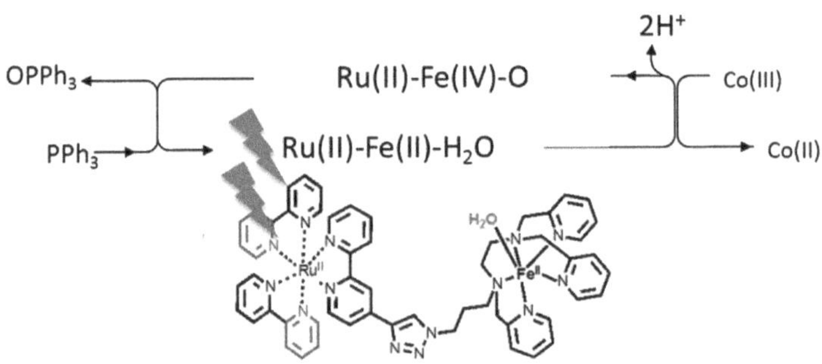

Figure 12 Photocatalytic OAT using a Ru-Fe dyad in the presence of water and $[Co(NH_3)_5Cl]^{2+}$ as sacrificial electron scavenger.

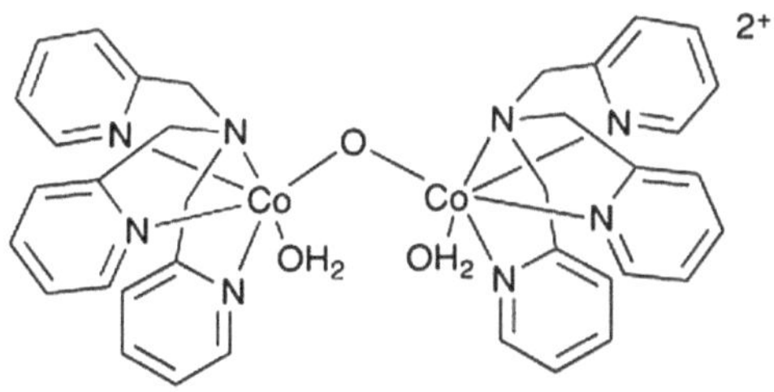

Figure 13 Dinuclear cobalt(II) complex as WOC.

was shown to act as a WO catalyst in the presence of a photoactive unit and a sacrificial EA. The topology of the two metal ions leads the authors to propose an intramolecular formation of the O—O bond through the radical-type coupling reaction of two oxo groups sitting on a cobalt(IV) species (Fig. 13).

Of course there has been much development in the chemistry of manganese complexes during these last decades. Our goal is not to give an exhaustive review on the state of art in this domain. The reader can refer to excellent reviews on this topic (Blakemore, Crabtree, & Brudvig, 2015; Kärkäs, et al., 2014; Limburg, Bouwman, & Bonnet, 2012; Najafpour et al., 2014; Yagi et al., 2009). We want to mention an interesting piece of work concerning the design of a biscompartmental anionic ligand that was used to prepare a dinuclear manganese(III) complex. The design of anionic ligand was again guided by the fact that anionic ligands will help to contain highly oxidizing manganese ions. Under irradiation and in the presence of $[Ru(bpy)_3]^{2+}$ and persulfate ions as sacrificial EA, O_2 was detected pertaining the photooxidation of water by the classic triptych mixture. On a mechanistic functioning of this system, it is argued that the dinuclear manganese complex rocks between Mn(III)Mn(III) and Mn(V) Mn(V) oxidation states (Karlsson et al., 2011) (Fig. 14).

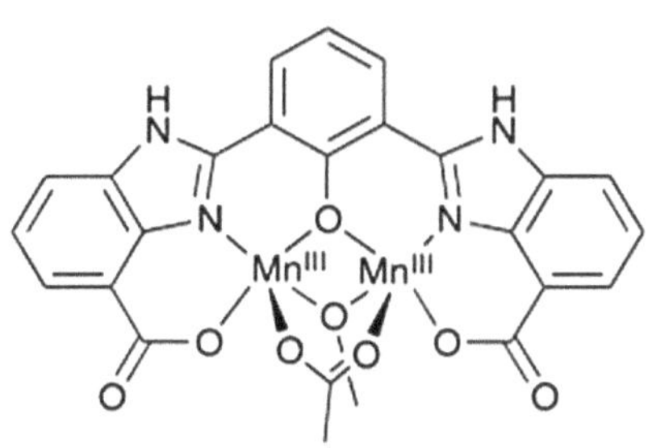

Figure 14 Akermark's dinuclear Mn(III) catalyst for WO.

5. MOLECULAR CATALYSTS FOR REDUCTION

This theme has known much success in comparison to the oxidative part of the water-splitting reaction. The underlying reason probably is that the two-electron catalytic process is less energetically demanding in comparison with the four-electron and four-proton processes for the water oxidation reaction. Molecular catalysts for the hydrogen-evolving reaction (HER) have been an active research domain for decades and have gained much impetus during these last years. Indeed, with the idea to develop a renewable energetic vector, hydrogen stands out as the favourite candidate. However, developing such an economy will rely on the progress in many areas such as its mass production, storage and transport. We mention here the contribution of molecular complexes in the production of H_2. The problem is to replace platinoid metals at the cathode of an electrolyser of the PEM type as we presented earlier. Many reviews have been published on this particular subject and we convey the interested readers to some of them (Artero, Chavarot-Kerlidou, & Fontecave, 2011; Dempsey, Brunschwig, Winkler, & Gray, 2009; Eckenhoff, McNamara, Du, & Eisenberg, 2013). Perhaps, the family of molecular complex that has made the most headlines in this field is the oximato complex family. The corresponding cobalt, nickel and iron derivatives have been studied and reported as efficient HER catalysts. The cobalt glyoximato complexes on their own have gathered particular attention probably because they were already known as HER catalysts and because their easy synthetic accessibility. Their electrochemical properties have been scrutinized under diverse experimental conditions and monometallic or bimetallic mechanisms were proposed for the formation of H_2. Several families of ligand containing the oximato group have been prepared and their electrocatalytic properties have been examined. Such a strategy has led to the fine-tuning of the overpotential and concomitantly reaching higher electrochemical performance for the production of H_2. Families of these complexes were used to elaborate membrane-electrode assemblies to replace platinum at the cathode of a PEM cell (Dinh Nguyen et al., 2012). Results from this study are quite promising and in Fig. 15 is represented the cell voltage during continuous electrolysis at a current density of 200 mA/cm^2. As we can notice the membrane-electrode assemblies with the cobalt complexes work at a higher potential for the same current density compared to the commercial one with platinum as the cathode. However, the current density was stable for days after runs of ON and OFF polarization. It is to be note that no insights on the nature of the catalytic species were given in this study and this is an important

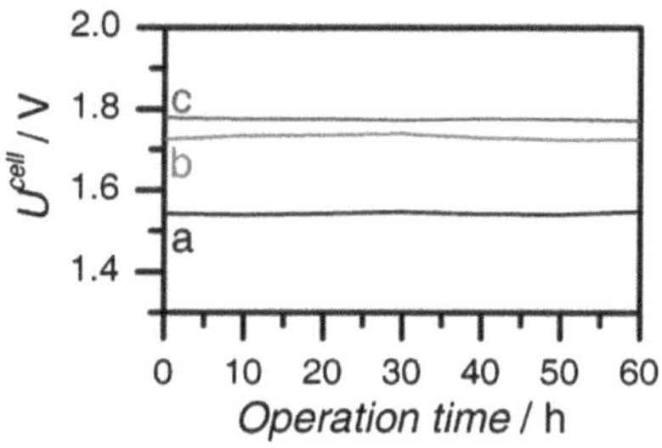

Figure 15 Change in cell voltage during continuous operation at 200 mA/cm^2. The cell was run on an average 7 h/day: a) Ir/Nafion117/Pt, b) Ir/Nafion117/[Co(dmg)$_3$(BF)$_2$]BF$_4$-VulcanXC72, c) Ir/Nafion117/[Co(dmgBF$_2$)$_2$]VulcanXC72. *Reprinted from Dinh Nguyen, M. T., Ranjbari, A., Catala, L., Brisset, F., Millet, P., & Aukauloo, A. (2012). Implementing molecular catalysts for hydrogen production in proton exchange membrane water electrolysers.* Coordination Chemistry Reviews, *256*(21–22), 2435–2444. with permission from Elsevier.

issue in declaring that the molecular complex at the surface of the graphite electrode is the actual catalytic species.

Serious interrogations were raised concerning a trisglyoximato-cobalt caged complex acting as an HER catalyst. Advanced electrochemical studies have evidenced that the molecular complex was just acting as a precursor for the formation of cobalt-based nanoparticles of ill-defined chemical formulae that were acting as the real catalyst. Subsequently more detailed electrochemical studies on the bisglyoximato complexes have, here too, shown the formation of similar cobalt-based nanoparticles at the surface of the graphite electrodes upon electrocatalysis (El Ghachtouli et al., 2013). These findings have again induced molecular chemists to take extra care to provide a flawless mechanistic route for the functioning of molecular catalysts (Artero & Fontecave, 2013). Interestingly, this chemical alteration process can be viewed as an alternative way to produce catalytic nanoparticles for the HER. Indeed, understanding the process of chemical alterations at the molecular level leading to efficient catalytic nanomaterials may open new routes to prepare cost-effective catalysts in this research field.

We must mention here the noticeable work of the Dubois group in the rational development of molecular catalysts for HER (Dubois & Dubois, 2009). The N- and P- containing ligands around a nickel ion lead to very efficient catalysts. The successful integration of second sphere participating ligands in the catalytic activity pattern clearly demonstrates the chemical tools at hand of molecular chemists to implement nature's chemical trick to enhance the catalytic properties of molecular complexes. These efficient molecular catalysts were successfully covalently attached to Carbon Nanotubes and their electrocatalytic activity was demonstrated (Le Goff et al., 2010). The modified electrode was found to have enhanced catalytic properties towards the production of H_2 working at low overpotential. More interestingly, the covalently linked catalyst was also capable of driving the oxidation of hydrogen. Although no real explanation was given for this observation, it is clear that molecular complexes in interaction with surfaces may lead to novel reactivity schemes. However, we will need to unravel the intimate interaction and the modification of the electronic properties of the metal complexes at the surface of different materials.

Concerning the photocatalytic production of H_2 with molecular complexes, the strategy has remained unchanged since decades. The classic assay consists of a mixture of chromophore, catalyst and sacrificial ED in the presence of a source of protons irradiated in the visible spectral window. The produced H_2 is monitored and characterized by GC—MS. Based on original work of Lehn and Sauvage concerning the photocatalytic production of H_2 with cobalt glyoximes complexes, several groups have used the same strategy to design more sophisticated assemblies of chromophores and different oximato complexes to increase the quantum yield of these triptych mixtures. Of course, in the continuing effort of molecular chemists to understand the functioning at a molecular level of the transfer of electrons and protons to produce H_2, much knowledge is being gathered. This constitutes the basic science that needs to be done prior to any technological deployment. Undoubtedly, developing molecular systems will require the coupling with surface chemistry in order to reach a solid device. Photocatalytic H_2 production has been realized by anchoring the photoactive unit and the catalytic module at the surface of TiO_2 particulates in the presence of a sacrificial ED (Lakadamyali & Reisner, 2011). This study therefore points to the feasibility of transferring molecular units on a semiconductor acting as an electron source for the photocatalytic H_2 production.

Unarguably, the research on bioinspired catalysts to perform the HER has known an unprecedented effort from the part of bioinorganic chemists. Since the first report of Rauchfuss and colleagues on the electrocatalytic production of H_2 using diiron complexes as models of the Iron-Hydrogenase catalytic site, a cornucopia of synthetic Fe—Fe and Ni—Fe complexes have been developed and the electrocatalytic properties disclosed. Undoubtedly, these models have served to understand the electronic and spectroscopic signatures of several intermediates in the proposed mechanistic schemes of the catalytic site of these enzymes. However, the electrocatalytic properties have been confined to quite negative potentials in comparison to the biological system. The inclusion of a sluggish synthetic model into a catalytically depleted iron-hydrogenase enzyme has proven the importance of the protein scaffold in conferring the unique catalytic properties of the Fe—Fe catalytic core. As such it seems that chemists still face the grand challenge to extract the right bits of information from natural catalysts to implement in a synthetic system to efficiently catalyse the production of H_2 with the most abundant transition element in the earth crust: iron.

Concomitantly to the research on the production of H_2, a similar effort is being addressed to the reduction of CO_2 (Appel et al., 2013; Morris, Meyer, & Fujita, 2009). Here too, noble-metal complexes share the top place for the reduction of CO_2 either by electrocatalysis or by photocatalysis. Ruthenium, rhenium and platinoid complexes show good efficiency. Recent findings using iron porphyrins to selectively reduce CO_2 to CO have been a major advancement in the field (Costentin, Passard, Robert, & Saveant, 2014). Well-known nickel cyclam derivatives have also been used as electrocatalysts at a glassy carbon electrode instead of mercury electrode. Likewise, manganese-based complexes have been reported as good electrocatalysts for the reduction of CO_2 to CO (Bourrez, Molton, Chardon-Noblat, & Deronzier, 2011). As discussed earlier, these examples must definitely be looked as the basis for the use of first-row transition-metal complexes in the design and fabrication of cheap catalysts. The grand problem will be to perform selective reduction of CO_2. For the time being, molecular chemists are addressing the two-electron, two-proton reduction of CO_2 either to CO or to HCOOH. Insights in the mechanistic routes to guide selectively these processes will be a stepping-stone in the control of multielectron reduction of CO_2.

6. DRIVING CATALYSTS BY LIGHT

As presented in the above section, most studies up to date use mixtures of chromophores and catalysts in homogenous solution (Artero et al., 2011; Parent & Sakai, 2014; Stoll et al., 2015). In this case the two components, chromophore and catalyst can be developed separately, and different combinations of chromophore/catalyst couples in different stoichiometric ratios can easily be studied. The drawback is that rates of intermolecular ET between the components are controlled by diffusion and it is therefore difficult to optimize rates and directionality of the ET.

Supramolecular systems where both chromophore and catalyst units are part of the same molecule and where the catalyst is activated by intramolecular ET are expected to allow better control of reaction pathways and rates. Many such systems have been synthesized and studied both for oxidation and reduction catalysis (Guillo et al., 2012; Herrero et al., 2010; Ladomenou et al., 2015; Li et al., 2012; Rocha et al., 2008). Their syntheses implying often long and difficult methods, a more efficient synthesis based on a click-chemistry approach has been proposed (Baron, Herrero, Quaranta, Charlot, Leibl, Aukauloo, et al., 2012; Baron, Herrero, Quaranta, Charlot, Leibl, Vauzeilles, et al., 2012; Herrero, Batchelor, et al., 2013). The bridging ligand between the chromophore and the catalyst units is of special importance. First, it should be rigid enough to keep the two units at a defined distance. Furthermore, it should provide the right electronic coupling, strong enough to permit efficient charge transfer between the modules, but not too strong to preserve their individual electronic properties.

The design of supramolecular assemblies is also pursued with the target to anchor these systems at the surface of electrodes to develop modified photoelectrodes. Prior to grafting they are usually first characterized in solution, where time-resolved spectroscopic studies are easier. Unfortunately, the commonly observed processes of energy transfer from the excited chromophore to metal ions of the catalyst in supramolecular complexes make these studies often more difficult. In general, this energy transfer is not productive as the catalyst excited state is rapidly deactivated by fast decay to the ground state. For this reason, in solution, supramolecular complexes might easily show lower performance for photocatalysis compared to chromophore–catalyst mixtures. However it is important to consider that this might no longer be the case once the supramolecular complexes are grafted to semiconductor electrodes, because of the ultrafast charge injection process which could outcompete kinetically intramolecular energy transfer quenching. Also, in some favourable cases, ping-pong-like energy transfer processes have been identified where fast back energy transfer from the catalyst triplet excited state to the chromophore triplet excited state leads to at least partial rescue of excitation energy (Lazarides et al., 2011; Quaranta et al., 2015; Yarnell, Deaton, McCusker, & Castellano, 2011).

The catalysts employed for photoactivation assays have usually been tested as electrocatalysts before. However, it should be considered that the mode of activation in the two cases could be different insofar as in photoactivation, electrons are added or removed from the catalyst one-by-one, whereas a metal electrode poised at the right potential could deliver more than one charge simultaneously. To give an example, some diiron proton reduction catalyst show a single two-electron reduction wave in cyclic voltammetry measurements due to the fact that the second reduction is easier than the first one (Felton et al., 2007; Gloaguen, Morvan, Capon, Schollhammer, & Talarmin, 2007). Thus, as a valuable complementary technique to electrochemical measurements, activation by flashes of light can give insight into the structure of intermediate species by advancing the reaction cycle electron by electron.

For the assay of photocatalytic activity in solution, most studies make use of irreversible, sacrificial EDs or acceptors added in large excess. The advantage is that back reactions are nearly completely avoided. As a consequence, the light-induced oxidation or reduction of the chromophore will occur with a quantum yield approaching unity, and the long life time of the activated chromophore species ensures efficient interaction with the catalysts even in chromophore–catalyst mixtures. The results therefore represent an upper limit for photocatalytic activity of a given system under real condition, that is, in the absence of sacrificial oxidants or reductants. However, as mentioned before, these sacrificial EAs and donors are noninnocent, and it is important to consider their possible contribution to light absorption and pH changes, as well as participation of their often very reactive degradation products in the catalytic cycle and in degradation of the molecular chromophores or catalysts.

7. CHALLENGES AND BOTTLENECKS

As a matter of fact, designing molecular sensitizer-catalyst ensembles to perform the light-driven multielectron catalytic transformation using water as electron and proton source is a major theme on the agenda of chemists. This highly competitive research field is currently being actively pursued in several labs on an international level, and the large number of publications clearly testifies of significant advances in the field. Nevertheless, they also show that significant challenges persist, all related to the requirements of efficiency, stability and cost, which often appear linked but mutually exclusive.

In a certain sense, the cost criterion seems the most easy to handle. It implies in particular avoidance of noble metals, although other aspects, such as costs for large-scale production, have also to be considered. For some noble metals it is not only their price which is the limiting factor, but the sheer quantities available on earth. Enzymes show that efficient alternatives are possible, but they are not very robust. Biological systems generally employ repair mechanisms to keep activity high, although at a not-negligible energetic cost (In the case of photosystem two one protein subunit is exchanged after about 10,000 turnovers, including recycling of pigments and reassembling of the Mn_4Ca cluster of the OEC.). Engineering self-healing in artificial constructs, however, is not yet realistic, although such aspects have been invoked in the context of catalyst self-assembly (Nocera, 2012). Formation of highly reactive intermediates cannot be avoided, especially for the water oxidation reaction. To increase stability, deleterious side reaction pathways have to be identified and minimized. This is a complex issue and implies characterization of structural changes of the photocatalysts induced under operating conditions. Also, here, immobilization of molecular complexes on electrodes can be expected to have beneficial effects by diminishing both potential intermolecular deactivation mechanisms and the degradation of highly charged intermediates.

Efficiency optimization is a similarly complex topic, involving light absorption and energetic and kinetic aspects of charge transfer. Fig. 16 shows the energetic constraints for an efficient supramolecular photocatalyst system for oxidation. The challenge is to adjust the potential of all redox couples within a narrow window to allow activation of the catalytic site without too much energy loss. This is particularly important for the four sequential oxidation steps of WOCs. In this context, proton-coupled ET (PCET) plays a crucial role as a mechanism which allows for charge compensation and avoids high-energy intermediates (Cukier & Nocera, 1998; Dogutan, McGuire, & Nocera, 2011; Huynh & Meyer, 2007); see also Fig. 17). Although energetic optimization translates into a small driving force for all ET steps, experimental observations tend to converge that this does not necessarily lead to slow ET kinetics.

Kinetic aspects are the main factors governing the quantum yield of the photocatalytic process. To keep the quantum yield high, it is important to minimize wasteful, but energetically very favourable back reactions, which risk annihilating charge accumulation on the catalytic site. This means to incorporate 'photodiode' functions in the system via kinetic optimization of ET steps, that is, favour ET reactions from the right to the left side in Fig. 16. A promising, bioinspired strategy is to increase the distance between the chromophore and the catalyst units and to introduce an electron relay function consisting of a redox-active group which in its ground state can be an ED but not an EA. The kinetics of reaction 2 which is in competition

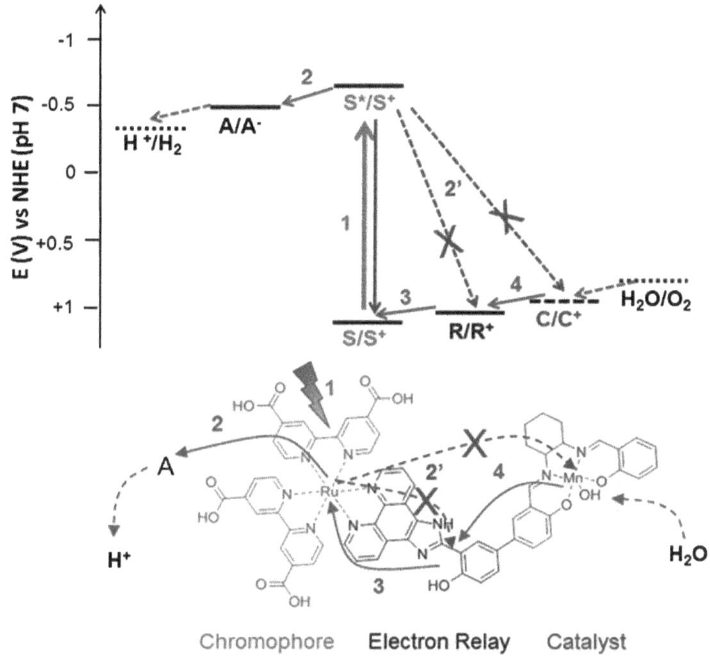

Figure 16 Simplified energetic scheme (top) and sequence of light-induced electron transfer reactions in a molecular photocatalyst for oxidation. Productive charge transfers are indicated by *green arrows* (grey in print versions), nonproductive ones by *red arrows* (dark grey in print versions). Possible back reactions from the reduced acceptor (A^-) to oxidized species (S^+, R^+, C^+) are omitted for clarity and because in most studies in solution these back reactions are suppressed by the use of irreversible EAs. *Reproduced with permission. ©(2015) SCF/L'Actualité Chimique.*

with reaction 2′ (Fig. 16) can be optimized by grafting of the chromophore onto semiconducting, n-type electrodes, such as TiO_2. In such systems, the electrode plays the role of a very efficient EA due to the ultrafast injection of electrons from the excited state of the chromophore into the conduction band of the semiconductor nanoparticles. In any case, the intrinsic decay time of the excited state of the chromophore sets a strict limit on the acceptable rate of the charge separation process, which should be at least one order of magnitude faster for high quantum yields.

For the development of molecular devices, the understanding of molecular mechanisms of light-activation and catalysis is of primordial importance. As an example, Herrero, Quaranta, Fallahpour, Leibl, and Aukauloo (2013) have recently shown how introduction of auxophores on the periphery of the chromophore in a chromophore–catalyst construct leads to a different and significantly improved mechanism of the catalyst activation.

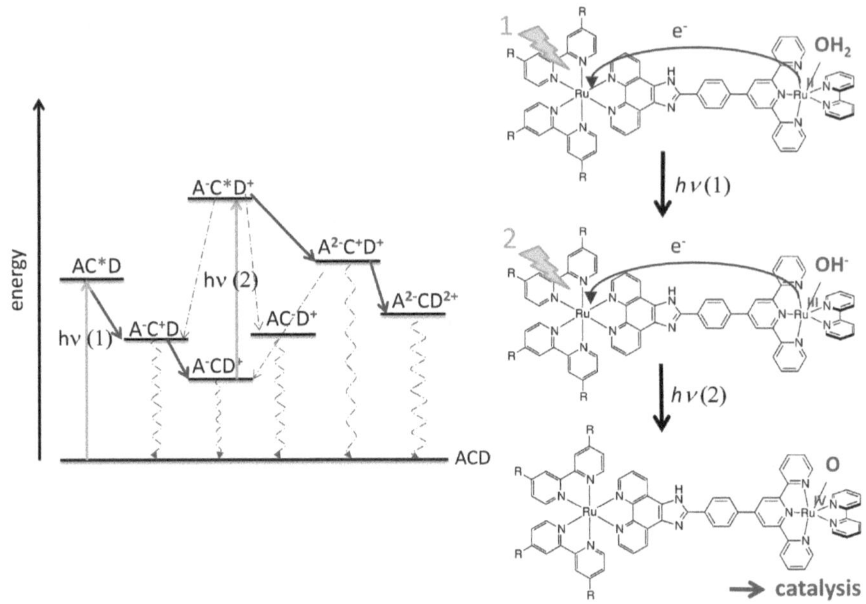

Figure 17 General reaction scheme for photoinduced electron transfer in an ACD system (A: electron acceptor, C: chromophore, D: electron donor). Possible charge transfer processes occurring upon one (1) and two (2) photon excitation. *Dotted arrows*: recombination reactions. *Bold blue arrows* (grey in print versions): reactions to optimize. EAs are omitted on the right side for clarity.

Another daunting challenge in the domain that would lead to breakthroughs is assuring efficient charge accumulation at the catalytic sites through vectorial electron transfer from/to a chromophore unit (Inoue et al., 2011). In other words, fighting all deleterious charge annihilation pathways (Fig. 17). This requires investigation of intramolecular electron transfer reactions beyond the one-electron process. Flash excitation allows adding or removing electrons one by one and to assess the yield of formation and the stability of photogenerated redox states at the catalytic sites. This kind of measurements is well established for reversible, single excitation events. In some cases, a one-electron activated intermediate can be prepared by chemical oxidation or reduction. But in the general case, double-flash experiments are needed to follow the second activation step and to determine the kinetic competition between useful forward and wasteful back electron transfer. Despite the great interest of such experiments, there are very few examples reported in the literature, probably due to the specific difficulties connected with them (Hammarström, 2015).

8. CONCLUDING REMARKS

Bioinspired photocatalysis for solar fuel production is an exciting interdisciplinary research field with essential input from biology, physics, chemistry and material science. It holds great promise due to the tunability of the molecular systems, but a great many bottlenecks still exist, especially with respect to efficiency and stability. Light absorption has to be optimized by developing stable sensitizers. Here, tandem configurations, where the oxidation and reduction part of the process harvests complementary ranges of the solar spectrum, should be aimed at. Efficient charge separation seems at reach with the use of semiconducting electrodes. But whereas TiO_2 seems established as suitable n-type semiconductor material for photoanodes, the equivalent material for photocathodes still represents a significant challenge (Odobel & Pellegrin, 2013). Concerning the difficulty to realize catalyst activation by efficient accumulation of charges from one-electron charge separation events, the use of electron relay functions was proposed as bioinspired feature. A clear demonstration of the benefit of such a function in supramolecular complexes was not yet reported.

Concerning the difficult point of WOC, we want to mention two very recent, encouraging results. A very close mimic of the OEC has been synthesized by C. Zhang et al. in 2015. Besides the Mn_4Ca-metal-oxygen, core this model complex included ligand groups with carboxylate functions closely mimicking the protein environment of the native OEC. Interestingly, the structural similarity between the original and the synthetic analogue extends to spectroscopic and electrochemical properties, but catalytic activity for water oxidation is not yet established. Further refinement, for example with respect to the binding sites for substrate water, is expected to bring valuable insight in structure–function relationship in this kind of systems.

The group of Llobet reported on a new family of molecular catalyst for water oxidation (Matheu, Ertem, et al., 2015). In electrocatalysis, one of these mononuclear ruthenium complexes showed impressive turnover frequencies of 8000 s^{-1} at neutral pH and 1.4 V/NHE and calculated turnover numbers in the range of 27 million. The latter value would satisfy the requirements for stability of 20 years in solar flux, whereas the high turnover frequency, very important in pure electrolyser applications, is about three orders of magnitude higher than what is required in a photocatalyst. If exchange of the ruthenium centre for a nonnoble metal such as Mn, Co or Fe could be traded in for 1000 times lower turnover frequency, by keeping the high stability, and decreasing the over potential by some 100 mV, the problem of WOC would be solved.

Finally, the field needs input from the theoretical field. The theory of PCET has received much attention and rightly so, but multielectron transfer pathways are poorly understood and have to be modelled by new approaches beyond classical Marcus theory (Lin, Balamurugan, Zhang, Skourtis, & Beratan, 2015) and also with respect to dynamic self-organization, and synergetic and positive feedback mechanisms (Tributsch, 1998). For the simulation and optimization of final PEC devices, multiscale modeling, which includes parameters from the molecular to the device level, are needed that permit to adequately account for the interactions of all phenomena (Kemppainen, Halme, & Lund, 2015).

REFERENCES

Alexy, E. J., Chandrashaker, V., & Lindsey, J. S. (2014). Perylene-tetrapyrrole arrays with panchromatic spectral coverage. *Abstracts of Papers of the American Chemical Society, 248.*

Appel, A. M., Bercaw, J. E., Bocarsly, A. B., Dobbek, H., DuBois, D. L., Dupuis, M., … Waldrop, G. L. (2013). Frontiers, opportunities, and challenges in biochemical and chemical catalysis of CO_2 Fixation. *Chemical Reviews, 113*(8), 6621–6658.

Archer, S., & Weinstein, J. A. (2012). Charge-separated excited states in platinum(II) chromophores: photophysics, formation, stabilization and utilization in solar energy conversion. *Coordination Chemistry Reviews, 256*(21–22), 2530–2561.

Artero, V., Chavarot-Kerlidou, M., & Fontecave, M. (2011). Splitting water with cobalt. *Angewandte Chemie International Edition, 50*(32), 7238–7266.

Artero, V., & Fontecave, M. (2013). Solar fuels generation and molecular systems: is it homogeneous or heterogeneous catalysis? *Chemical Society Reviews, 42*(6), 2338–2356.

Ashford, D. L., Gish, M. K., Vannucci, A. K., Brennaman, M. K., Templeton, J. L., Papanikolas, J. M., & Meyer, T. J. (2015). Molecular chromophore-catalyst assemblies for solar fuel applications. *Chemical Reviews, 115*(23), 13006–13049.

Bae, E. Y., Choi, W. Y., Park, J. W., Shin, H. S., Kim, S. B., & Lee, J. S. (2004). Effects of surface anchoring groups (carboxylate vs phosphonate) in ruthenium-complex-sensitized TiO_2 on visible light reactivity in aqueous suspensions. *Journal of Physical Chemistry B, 108*(37), 14093–14101.

Baron, A., Herrero, C., Quaranta, A., Charlot, M. F., Leibl, W., Aukauloo, A., & Vauzeilles, B. (2012a). Click-triazole as an efficient connector for Ru(II) bipyridine-based devices. *Abstracts of Papers of the American Chemical Society*, 243.

Baron, A., Herrero, C., Quaranta, A., Charlot, M. F., Leibl, W., Vauzeilles, B., & Aukauloo, A. (2012b). Click chemistry on a ruthenium polypyridine complex. An efficient and versatile synthetic route for the synthesis of photoactive modular assemblies. *Inorganic Chemistry, 51*(11), 5985–5987.

Bella, F., Gerbaldi, C., Barolo, C., & Gratzel, M. (2015). Aqueous dye-sensitized solar cells. *Chemical Society Reviews, 44*(11), 3431–3473.

Berardi, S., Drouet, S., Francas, L., Gimbert-Surinach, C., Guttentag, M., Richmond, C., … Llobet, A. (2014). Molecular artificial photosynthesis. *Chemical Society Reviews, 43*(22), 7501–7519.

Blakemore, J. D., Crabtree, R. H., & Brudvig, G. W. (2015). Molecular catalysts for water oxidation. *Chemical Reviews, 115*(23), 12974–13005, 150720112150009.

Bourrez, M., Molton, F., Chardon-Noblat, S., & Deronzier, A. (2011). [Mn(bipyridyl)$(CO)_3Br$]: an abundant metal carbonyl complex as efficient electrocatalyst for CO_2 reduction. *Angewandte Chemie International Edition, 50*(42), 9903–9906.

Bozoglian, F., Romain, S., Ertem, M. Z., Todorova, T. K., Sens, C., Mola, J., ... Llobet, A. (2009). The Ru-Hbpp water oxidation catalyst. *Journal of the American Chemical Society, 131*(42), 15176–15187.

Castellano, F. N. (2015). Altering molecular photophysics by merging organic and inorganic chromophores. *Accounts of Chemical Research, 48*(3), 828–839.

Chen, G., Chen, L. J., Ng, S. M., Man, W. L., & Lau, T. C. (2013). Chemical and visible-light-driven water oxidation by iron complexes at pH 7–9: evidence for dual-active intermediates in iron-catalyzed water oxidation. *Angewandte Chemie International Edition, 52*(6), 1789–1791.

Chen, Q., Montesarchio, D., & Hellingwerf, K. J. (2016). 'Direct Conversion': Artificial Photosynthesis with Cyanobacteria. In R. Bruno (Ed.). *Artificial Photosynthesis* (Vol. 79, pp. 43–62).

Chen, H. Y., Tagore, R., Das, S., Incarvito, C., Faller, J. W., Crabtree, R. H., ... Brudvig, G. W. (2005). General synthesis of di-μ-oxo dimanganese complexes as functional models for the oxygen evolving complex of photosystem II. *Inorganic Chemistry, 44*(21), 7661–7670.

Collins, T. J. (2002). TAML oxidant activators: a new approach to the activation of hydrogen peroxide for environmentally significant problems. *Accounts of Chemical Research, 35*(9), 782–790.

Concepcion, J. J., Jurss, J. W., Brennaman, M. K., Hoertz, P. G., Patrocinio, A. O. T., Iha, N. Y. M., ... Meyer, T. J. (2009). Making oxygen with ruthenium complexes. *Accounts of Chemical Research, 42*(12), 1954–1965.

Concepcion, J. J., Jurss, J. W., Norris, M. R., Chen, Z. F., Templeton, J. L., & Meyer, T. J. (2010). Catalytic water oxidation by single-site ruthenium catalysts. *Inorganic Chemistry, 49*(4), 1277–1279.

Concepcion, J. J., Jurss, J. W., Templeton, J. L., & Meyer, T. J. (2008). One site is enough. Catalytic water oxidation by $[Ru(tpy)(bpm)(OH_2)]^{(2+)}$ and $[Ru(tpy)(bpz)(OH_2)]^{(2+)}$. *Journal of the American Chemical Society, 130*(49), 16462–16463.

Concepcion, J. J., Tsai, M. K., Muckerman, J. T., & Meyer, T. J. (2010). Mechanism of water oxidation by single-site ruthenium complex catalysts. *Journal of the American Chemical Society, 132*(5), 1545–1557.

Coridan, R. H., Nielander, A. C., Francis, S. A., McDowell, M. T., Dix, V., Chatman, S. M., & Lewis, N. S. (2015). Methods for comparing the performance of energy-conversion systems for use in solar fuels and solar electricity generation. *Energy and Environmental Science, 8*(10), 2886–2901.

Costentin, C., Passard, G., Robert, M., & Saveant, J. M. (2014). Ultraefficient homogeneous catalyst for the CO_2-to-CO electrochemical conversion. *Proceedings of the National Academy of Sciences of the United States of America, 111*(42), 14990–14994.

Cukier, R. I., & Nocera, D. G. (1998). Proton-coupled electron transfer. *Annual Review of Physical Chemistry, 49*, 337–369.

Dempsey, J. L., Brunschwig, B. S., Winkler, J. R., & Gray, H. B. (2009). Hydrogen evolution catalyzed by cobaloximes. *Accounts of Chemical Research, 42*(12), 1995–2004.

Dinh Nguyen, M. T., Ranjbari, A., Catala, L., Brisset, F., Millet, P., & Aukauloo, A. (2012). Implementing molecular catalysts for hydrogen production in proton exchange membrane water electrolysers. *Coordination Chemistry Reviews, 256*(21–22), 2435–2444.

Dismukes, G. C., Brimblecombe, R., Felton, G. A. N., Pryadun, R. S., Sheats, J. E., Spiccia, L., & Swiegers, G. F. (2009). Development of bioinspired Mn_4O_4-Cubane water oxidation catalysts: lessons from photosynthesis. *Accounts of Chemical Research, 42*(12), 1935–1943.

Dogutan, D. K., McGuire, R., & Nocera, D. G. (2011). Electocatalytic water oxidation by cobalt(III) hangman β-octafluoro corroles. *Journal of the American Chemical Society, 133*(24), 9178–9180.

Dubois, M. R., & Dubois, D. L. (2009). Development of molecular electrocatalysts for CO_2 reduction and H_2 production/oxidation. *Accounts of Chemical Research, 42*(12), 1974–1982.

Eckenhoff, W. T., McNamara, W. R., Du, P., & Eisenberg, R. (2013). Cobalt complexes as artificial hydrogenases for the reductive side of water splitting. *Biochimica et Biophysica Acta (BBA) – Bioenergetics, 1827*(8–9), 958–973.

El Ghachtouli, S., Fournier, M., Cherdo, S., Guillot, R., Charlot, M.-F., Anxolabéhère-Mallart, E., … Aukauloo, A. (2013). Monometallic cobalt–trisglyoximato complexes as precatalysts for catalytic H_2 Evolution in water. *Journal of Physical Chemistry C, 117*(33), 17073–17077.

Fakharuddin, A., Jose, R., Brown, T. M., Fabregat-Santiago, F., & Bisquert, J. (2014). A perspective on the production of dye-sensitized solar modules. *Energy and Environmental Science,* 7(12), 3952–3981.

Fan, K., Li, F. S., Wang, L., Daniel, Q., Gabrielsson, E., & Sun, L. C. (2014). Pt-free tandem molecular photoelectrochemical cells for water splitting driven by visible light. *Physical Chemistry Chemical Physics, 16*(46), 25234–25240.

Felton, G. A. N., Vannucci, A. K., Chen, J. Z., Lockett, L. T., Okumura, N., Petro, B. J., … Lichtenberger, D. L. (2007). Hydrogen generation from weak acids: electrochemical and computational studies of a diiron hydrogenase mimic. *Journal of the American Chemical Society, 129*(41), 12521–12530.

Funyu, S., Isobe, T., Takagi, S., Tryk, D. A., & Inoue, H. (2003). Highly efficient and selective epoxidation of alkenes by photochemical oxygenation sensitized by a ruthenium(II) porphyrin with water as both electron and oxygen donor. *Journal of the American Chemical Society, 125*(19), 5734–5740.

Gärtner, F., Cozzula, D., Losse, S., Boddien, A., Anilkumar, G., Junge, H., … Beller, M. (2011). Synthesis, characterisation and application of iridium(III) photosensitisers for catalytic water reduction. *Chemistry – A European Journal, 17*(25), 6998–7006.

Gloaguen, F., Morvan, D., Capon, J.-F., Schollhammer, P., & Talarmin, J. (2007). Electrochemical proton reduction at mild potentials by monosubstituted diiron organometallic complexes bearing a benzenedithiolate bridge. *Journal of Electroanalytical Chemistry, 603*(1), 15–20.

Guillo, P., Hamelin, O., Batat, P., Jonusauskas, G., McClenaghan, N. D., & Menage, S. (2012). Photocatalyzed sulfide oxygenation with water as the unique oxygen atom source. *Inorganic Chemistry, 51*(4), 2222–2230.

Hammarström, L. (2015). Accumulative charge separation for solar fuels production: coupling light-induced single electron transfer to multielectron catalysis. *Accounts of Chemical Research, 48*(3), 840–850.

Han, Z., Qiu, F., Eisenberg, R., Holland, P. L., & Krauss, T. D. (2012). Robust photogeneration of H_2 in water using semiconductor nanocrystals and a nickel catalyst. *Science, 338*(6112), 1321–1324.

Hanson, K., Brennaman, M. K., Luo, H., Glasson, C. R. K., Concepcion, J. J., Song, W., & Meyer, T. J. (2012). Photostability of phosphonate-derivatized, Ru^{II} polypyridyl complexes on metal oxide surfaces. *ACS Applied Materials and Interfaces, 4*(3), 1462–1469.

Hardin, B. E., Snaith, H. J., & McGehee, M. D. (2012). The renaissance of dye-sensitized solar cells. *Nature Photonics, 6*(3), 162–169.

Harlang, T. C., Liu, Y., Gordivska, O., Fredin, L. A., Ponseca, C. S., Jr., Huang, P., … Wärnmark, K. (2015). Iron sensitizer converts light to electrons with 92% yield. *Nature Chemistry,* 7(11), 883–889.

Herrero, C., Batchelor, L., Baron, A., El Ghachtouli, S., Sheth, S., Guillot, R., … Aukauloo, A. (2013). Click chemistry as a convenient tool for the incorporation of a ruthenium chromophore and a nickel-salen monomer into a visible-light-active assembly. *European Journal of Inorganic Chemistry*, (4), 494–499.

Herrero, C., Hughes, J. L., Quaranta, A., Cox, N., Rutherford, A. W., Leibl, W., & Aukauloo, A. (2010). Intramolecular light induced activation of a Salen-Mn-III complex by a ruthenium photosensitizer. *Chemical Communications, 46*(40), 7605–7607.

Herrero, C., Quaranta, A., Fallahpour, R. A., Leibl, W., & Aukauloo, A. (2013). Identification of the different mechanisms of activation of a [Ru-II(tpy)(bpy)(OH_2)]$^{(2+)}$ catalyst by modified ruthenium sensitizers in supramolecular complexes. *Journal of Physical Chemistry C, 117*(19), 9605–9612.

Herrero, C., Quaranta, A., Sircoglou, M., Senechal-David, K., Baron, A., Marin, I. M., … Banse, F. (2015). Successive light-induced two electron transfers in a Ru-Fe supramolecular assembly: from Ru-Fe(II)-OH_2 to Ru-Fe(IV)-oxo. *Chemical Science, 6*(4), 2323–2327.

Huynh, M. H., & Meyer, T. J. (2007). Proton-coupled electron transfer. *Chemical Reviews, 107*(11), 5004–5064.

Iali, W., Lanoe, P. H., Torelli, S., Jouvenot, D., Loiseau, F., Lebrun, C., … Ménage, S. (2015). A ruthenium(II)-copper(II) dyad for the photocatalytic oxygenation of organic substrates mediated by dioxygen activation. *Angewandte Chemie International Edition, 54*(29), 8415–8419.

Inoue, H., Shimada, T., Kou, Y., Nabetani, Y., Masui, D., Takagi, S., & Tachibana, H. (2011). The water oxidation bottleneck in artificial photosynthesis: how can we get through it? an alternative route involving a two-electron process. *ChemSusChem*, 173–179.

Jacobsson, T. J., Fjallstrom, V., Edoff, M., & Edvinsson, T. (2014). Sustainable solar hydrogen production: from photoelectrochemical cells to PV-electrolyzers and back again. *Energy and Environmental Science, 7*(7), 2056–2070.

Juris, A., Balzani, V., Barigelletti, F., Campagna, S., Belser, P., & Vonzelewsky, A. (1988). Ru(II) polypyridine complexes – photophysics, photochemistry, electrochemistry, and chemi-luminescence. *Coordination Chemistry Reviews, 84*, 85–277.

Kalyanasundaram, K. (1982). Photophysics, photochemistry and solar-energy conversion with tris(bipyridyl)ruthenium(II) and its analogs. *Coordination Chemistry Reviews, 46*(Oct), 159–244.

Kärkäs, M. D., Verho, O., Johnston, E. V., & Åkermark, B. (2014). Artificial photosynthesis: molecular systems for catalytic water oxidation. *Chemical Reviews, 114*(24), 11863–12001.

Karlsson, E. A., Lee, B. L., Akermark, T., Johnston, E. V., Karkas, M. D., Sun, J. L., … Åkermark, B. (2011). Photosensitized water oxidation by use of a bioinspired manganese catalyst. *Angewandte Chemie International Edition, 50*(49), 11715–11718.

Kemppainen, E., Halme, J., & Lund, P. (2015). Physical modeling of photoelectrochemical hydrogen production devices. *Journal of Physical Chemistry C, 119*(38), 21747–21766.

Kodis, G., Terazono, Y., Liddell, P. A., Andreasson, J., Garg, V., Hambourger, M., … Gust, D. (2006). Energy and photoinduced electron transfer in a wheel-shaped artificial photosynthetic antenna-reaction center complex. *Journal of the American Chemical Society, 128*(6), 1818–1827.

Ladomenou, K., Natali, M., Iengo, E., Charalampidis, G., Scandola, F., & Coutsolelos, A. G. (2015). Photochemical hydrogen generation with porphyrin-based systems. *Coordination Chemistry Reviews, 304–305*, 38–54.

Lakadamyali, F., & Reisner, E. (2011). Photocatalytic H_2 evolution from neutral water with a molecular cobalt catalyst on a dye-sensitised TiO_2 nanoparticle. *Chemical Communications, 47*(6), 1695.

Lazarides, T., McCormick, T., Du, P. W., Luo, G. G., Lindley, B., & Eisenberg, R. (2009). Making hydrogen from water using a homogeneous system without noble metals. *Journal of the American Chemical Society, 131*(26), 9192–+.

Lazarides, T., McCormick, T. M., Wilson, K. C., Lee, S., McCamant, D. W., & Eisenberg, R. (2011). Sensitizing the sensitizer: the synthesis and photophysical study of bodipy-Pt(II)(diimine)(dithiolate) conjugates. *Journal of the American Chemical Society, 133*(2), 350–364.

Le Goff, A., Artero, V., Metayé, R., Moggia, F., Jousselme, B., Razavet, M., … Fontecave, M. (2010). Immobilization of FeFe hydrogenase mimics onto carbon and gold electrodes by controlled aryldiazonium salt reduction: an electrochemical, XPS and ATR-IR study. *International Journal of Hydrogen Energy, 35*(19), 10790–10796.

Li, F., Fan, K., Xu, B., Gabrielsson, E., Daniel, Q., Li, L., & Sun, L. (2015). Organic dye-sensitized tandem photoelectrochemical cell for light driven total water splitting. *Journal of the American Chemical Society, 137*(28), 9153–9159.

Li, F., Jiang, Y., Zhang, B. B., Huang, F., Gao, Y., & Sun, L. C. (2012). Towards a solar fuel device: light-driven water oxidation catalyzed by a supramolecular assembly. *Angewandte Chemie International Edition, 51*(10), 2417–2420.

Limburg, J. (1999). A functional model for O–O bond formation by the O_2-evolving complex in photosystem II. *Science, 283*(5407), 1524–1527.

Limburg, B., Bouwman, E., & Bonnet, S. (2012). Molecular water oxidation catalysts based on transition metals and their decomposition pathways. *Coordination Chemistry Reviews, 256*(15–16), 1451–1467.

Lin, J., Balamurugan, D., Zhang, P., Skourtis, S. S., & Beratan, D. N. (2015). Two-electron transfer pathways. *Journal of Physical Chemistry B, 119*(24), 7589–7597.

Lü, J., Sheahan, C., & Fu, P. (2011). Metabolic engineering of algae for fourth generation biofuels production. *Energy and Environmental Science, 4*(7), 2451.

Matheu, R., Ertem, M. Z., Benet-Buchholz, J., Coronado, E., Batista, V. S., Sala, X., & Llobet, A. (2015). Intramolecular proton transfer boosts water oxidation catalyzed by a Ru complex. *Journal of the American Chemical Society, 137*(33), 10786–10795, 150812101708002.

Matheu, R., Francàs, L., Chernev, P., Ertem, M. Z., Batista, V., Haumann, M., … Llobet, A. (2015). Behavior of the Ru-bda water oxidation catalyst covalently anchored on glassy carbon electrodes. *ACS Catalysis, 5*(6), 3422–3429.

Mbemba Kiele, N., Herrero, C., Ranjbari, A., Aukauloo, A., Grigoriev, S. A., Villagra, A., & Millet, P. (2013). Ruthenium-based molecular compounds for oxygen evolution in acidic media. *International Journal of Hydrogen Energy, 38*(20), 8590–8596.

McKone, J. R., Lewis, N. S., & Gray, H. B. (2014). Will solar-driven water-splitting devices see the light of day? *Chemistry of Materials, 26*(1), 407–414.

Morris, A. J., Meyer, G. J., & Fujita, E. (2009). Molecular approaches to the photocatalytic reduction of carbon dioxide for solar fuels. *Accounts of Chemical Research, 42*(12), 1983–1994.

Najafpour, M. M., Isaloo, M. A., Eaton-Rye, J. J., Tomo, T., Nishihara, H., Satoh, K., … Allakhverdiev, S. I. (2014). Water exchange in manganese-based water-oxidizing catalysts in photosynthetic systems: from the water-oxidizing complex in photosystem II to nano-sized manganese oxides. *Biochimica et Biophysica Acta (BBA) – Bioenergetics, 1837*(9), 1395–1410.

Nocera, D. G. (2012). The artificial leaf. *Accounts of Chemical Research, 45*(5), 767–776.

Odobel, F., & Pellegrin, Y. (2013). Recent advances in the sensitization of wide-band-gap nanostructured p-type semiconductors. Photovoltaic and photocatalytic applications. *Journal of Physical Chemistry Letters, 4*(15), 2551–2564.

Oregan, B., & Gratzel, M. (1991). A low-cost, high-efficiency solar-cell based on dye-sensitized colloidal TiO_2 films. *Nature, 353*(6346), 737–740.

Panda, C., Debgupta, J., Diaz, D. D., Singh, K. K., Gupta, S. S., & Dhar, B. B. (2014). Homogeneous photochemical water oxidation by biuret-modified Fe-TAML: evidence of Fe-V(O) intermediate. *Journal of the American Chemical Society, 136*(35), 12273–12282.

Panda, M. K., Ladomenou, K., & Coutsolelos, A. G. (2012). Porphyrins in bio-inspired transformations: light-harvesting to solar cell. *Coordination Chemistry Reviews, 256*(21–22), 2601–2627.

Parent, A. R., & Sakai, K. (2014). Progress in base-metal water oxidation catalysis. *ChemSusChem, 7*(8), 2070–2080.

Quaranta, A., Charalambidis, G., Herrero, C., Margiola, S., Leibl, W., Coutsolelos, A., & Aukauloo, A. (2015). Synergistic "ping-pong" energy transfer for efficient light activation in a chromophore-catalyst dyad. *Physical Chemistry Chemical Physics, 17*(37), 24166–24172.

Rocha, R. C., Rein, F. N., Jude, H., Shreve, A. P., Concepcion, J. J., & Meyer, T. J. (2008). Observation of three intervalence-transfer bands for a class II-III mixed-valence complex of ruthenium. *Angewandte Chemie International Edition, 47*(3), 503–506.

Rodriguez, C. A., Modestino, M. A., Psaltis, D., & Moser, C. (2014). Design and cost considerations for practical solar-hydrogen generators. *Energy and Environmental Science, 7*(12), 3828–3835.

Romero, I., Rodriguez, M., Sens, C., Mola, J., Kollipara, M. R., Francas, L., ... Llobet, A. (2008). Ru complexes that can catalytically oxidize water to molecular dioxygen. *Inorganic Chemistry, 47*(6), 1824–1834.

Rotzinger, F. P., Munavalli, S., Comte, P., Hurst, J. K., Gratzel, M., Pern, F. J., & Frank, A. J. (1987). A molecular water-oxidation catalyst derived from ruthenium diaqua bis(2,2′-bipyridyl-5,5′-dicarboxylic acid). *Journal of the American Chemical Society, 109*, 6619–6626.

Ruberu, T. P. A., Dong, Y., Das, A., & Eisenberg, R. (2015). Photoelectrochemical generation of hydrogen from water using a CdSe quantum dot-sensitized photocathode. *ACS Catalysis, 5*(4), 2255–2259.

Sala, X., Romero, I., Rodriguez, M., Escriche, L., & Llobet, A. (2009). Molecular catalysts that oxidize water to dioxygen. *Angewandte Chemie International Edition, 48*(16), 2842–2852.

Sato, S., Arai, T., & Morikawa, T. (2015). Toward solar-driven photocatalytic CO_2 reduction using water as an electron donor. *Inorganic Chemistry, 54*(11), 5105–5113.

Song, W., Ito, A., Binstead, R. A., Hanson, K., Luo, H., Brennaman, M. K., ... Meyer, T. J. (2013). Accumulation of multiple oxidative equivalents at a single site by cross-surface electron transfer on TiO_2. *Journal of the American Chemical Society, 135*(31), 11587–11594.

Stoll, T., Castillo, C. E., Kayanuma, M., Sandroni, M., Daniel, C., Odobel, F., ... Collomba, M.-N. (2015). Photo-induced redox catalysis for proton reduction to hydrogen with homogeneous molecular systems using rhodium-based catalysts. *Coordination Chemistry Reviews, 304-305*, 20–37.

Stracke, J. J., & Finke, R. G. (2011). Electrocatalytic water oxidation beginning with the cobalt polyoxometalate $[Co_4(H_2O)_2(PW_9O_{34})_2]^{10-}$: identification of heterogeneous CoO_x as the dominant catalyst. *Journal of the American Chemical Society, 133*(38), 14872–14875.

Suseno, S., McCrory, C. C. L., Tran, R., Gul, S., Yano, J., & Agapie, T. (2015). Molecular mixed-metal manganese oxido cubanes as precursors to heterogeneous oxygen evolution catalysts. *Chemistry – A European Journal, 21*(38), 13420–13430.

Swierk, J. R., & Mallouk, T. E. (2013). Design and development of photoanodes for water-splitting dye-sensitized photoelectrochemical cells. *Chemical Society Reviews, 42*(6), 2357–2387.

Tributsch, H. (1998). Electron transfer: classical approaches and new frontiers. *Science, 279*(5358), 1891–1895.

Wang, P., Klein, C., Moser, J. E., Humphry-Baker, R., Cevey-Ha, N. L., Charvet, R., ... Grätzel, M. (2004). Amphiphilic ruthenium sensitizer with 4,4′-diphosphonic acid-2,2′-bipyridine as anchoring ligand for nanocrystalline dye sensitized solar cells. *Journal of Physical Chemistry B, 108*(45), 17553–17559.

Wasylenko, D. J., Palmer, R. D., Schott, E., & Berlinguette, C. P. (2012). Interrogation of electrocatalytic water oxidation mediated by a cobalt complex. *Chemical Communications, 48*(15), 2107–2109.

Wenger, O. S. (2009). Long-range electron transfer in artificial systems with d^6 and d^8 metal photosensitizers. *Coordination Chemistry Reviews, 253*(9–10), 1439–1457.

Yagi, M., Syouji, A., Yamada, S., Komi, M., Yamazaki, H., & Tajima, S. (2009). Molecular catalysts for wateroxidation toward artificial photosynthesis. *Photochemical and Photobiological Sciences, 8*(2), 139–147.

Yarnell, J. E., Deaton, J. C., McCusker, C. E., & Castellano, F. N. (2011). Bidirectional "ping-pong" energy transfer and 3000-fold lifetime enhancement in a Re(I) charge transfer complex. *Inorganic Chemistry, 50*(16), 7820–7830.

Young, K. J., Brennan, B. J., Tagore, R., & Brudvig, G. W. (2015). Photosynthetic water oxidation: insights from manganese model chemistry. *Accounts of Chemical Research, 48*(3), 567–574.

Zhang, C., Chen, C., Dong, H., Shen, J. R., Dau, H., & Zhao, J. (2015). A synthetic Mn_4Ca-cluster mimicking the oxygen-evolving center of photosynthesis. *Science, 348*(6235), 690–693.

Zhang, P., Wang, M., Dong, J. F., Li, X. Q., Wang, F., Wu, L. Z., & Sun, L. (2010). Photocatalytic hydrogen production from water by noble-metal-free molecular catalyst systems containing Rose Bengal and the cobaloximes of BFx-bridged oxime ligands. *Journal of Physical Chemistry C, 114*(37), 15868–15874.

Zheng, B., Sabatini, R. P., Fu, W.-F., Eum, M.-S., Brennessel, W. W., Wang, L., … Eisenberg, R. (2015). Light-driven generation of hydrogen: new chromophore dyads for increased activity based on bodipy dye and Pt(diimine)(dithiolate) complexes. *Proceedings of the National Academy of Sciences of the United States of America, 112*(30), E3987–E3996.

CHAPTER FOUR

Artificial Photosynthesis – An Inorganic Approach

S. Fiechter
Helmholtz-Zentrum Berlin für Materialien und Energie GmbH, Berlin, Germany
E-mail: fiechter@helmholtz-berlin.de

Contents

Abstract

The development of energy converting devices, which are able to light-induced split water into hydrogen and oxygen, has attracted increasing notice in the last two decades leading to numerous research activities worldwide addressing fundamental questions in the better understanding of photosynthesis as well as in the design of bioreactors and inorganic systems producing hydrogen from sunlight. One approach is the development of photoelectrodes consisting of photoactive semiconductor layers decorated with appropriate catalysts on top which immersed in an aqueous electrolyte and illuminated by sunlight can produce hydrogen. The high interest in all types of 'artificial leafs' is caused by the awareness that they offer attractive routes to convert sunlight into chemical energy storable in large quantities with the option to replace fossil fuels by renewable hydrogen gas and/or hydrocarbons. This plays an important role having the warming of the earth atmosphere by emission of CO_2 in mind. It is well known that more than 80% of the energy demand of mankind is based on chemical energy conversion, presently realized

Advances in Botanical Research, Volume 79
ISSN 0065-2296
http://dx.doi.org/10.1016/bs.abr.2016.04.002

by the burning of fossil fuels. For this, a changeover to a sustainable energy supply by regenerative hydrogen producing energy cycles is urgently requested based on the conversion of nearly infinite available sunlight into chemical energy.

1. INTRODUCTION

The energy supply in the future will be a decentralized one. Renewable electrical as well as chemical energy will be produced in part on roofs and at the facades of buildings by solar cells combined with light-driven hybrid electrolysers during daytime and the energy produced partly stored in batteries and gas tanks to keep them available at night or on days where the sun is shaded by clouds. Fig. 1 is showing such a scenario of a city, exclusively depending on solar hydrogen. Here it should be kept in mind that during daytime, electrical energy produced by solar cells should directly be used avoiding losses by further conversion processes. Presently, ecologically produced energy will be harvested from solar cells or can be gained from combined heat and power plants burning (bio)diesel or (bio)methane. However, in the present stage of energy transition from fossil fuels and nuclear power plants to renewables, security of supply is still depending on power plants burning fossil fuels, such as brown and black coal (Fig. 2).

In 2010, the energy demand of mankind amounted to 10^{16} TW per annum. This enormous quantity of energy is equivalent to the energy of the sunlight irradiated to earth's surface within 1 h showing that solar radiation is in principle able to cover our energy need. Even in 1912, Ciamician pointed in a visionary article published in Science to the fact that the energy from the sun is succeeding by orders of magnitude all fossil fuels extracted during one year. 'For our purposes the fundamental problem from the technical point of view is how to fix the solar energy through photochemical reactions' (Ciamician, 1912). Even 100 years later this question is still open documented by the high dependence of industrial countries from fossil fuels. An energy flow diagnostics of the United States from 2013 is impressively illustrating this addiction. In that year, only 1.2% of the energy to produce electricity was delivered by renewable energies in the United States (The use of self-generated electricity is not included in the diagram). The flow diagnostics also demonstrate that the energy efficiency only amounted to 59% showing the enormous potential to safe energy by an increase of the efficiencies of all kinds of processes (Lawrence Livermore National Laboratory; see also Fig. 2).

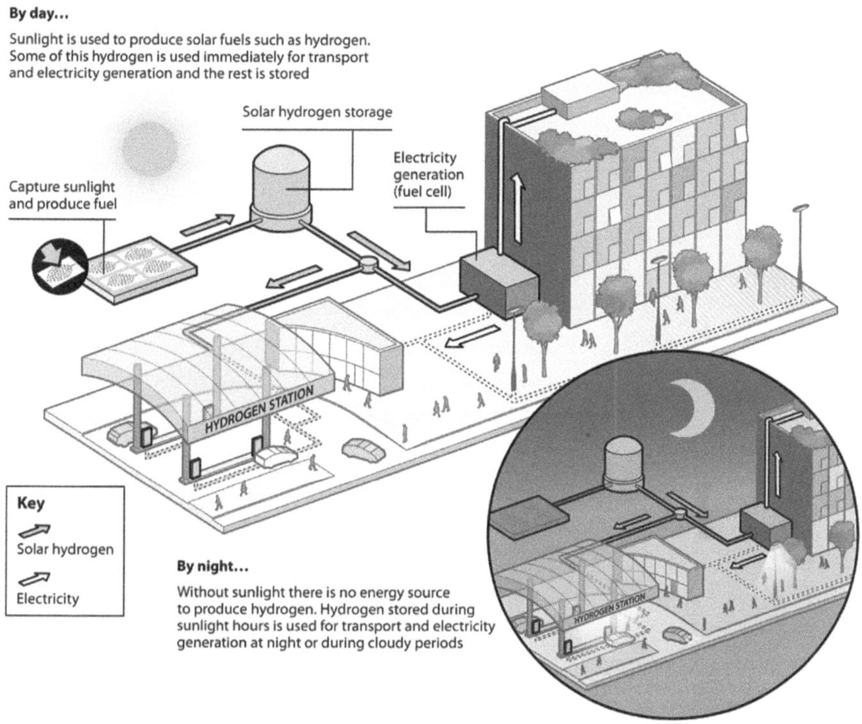

Figure 1 Scheme of a city exclusively powered by hydrogen (Barber, 2012). While solar energy is converted into hydrogen during daytime using artificial leaf-type structures, it will be reconverted into electricity during the night. The chemical energy produced can be used to fuel cars all the time. *Reproduced by permission of The Royal Society of Chemistry from Solar Fuels and Artificial Photosynthesis, January 2012, http://www.rsc.org/solar-fuels.*

In Germany, where wind and solar power have been remarkably enhanced in the last years, peak load energies of 38.5 GW in photovoltaics and of 35.3 GW in wind were contributing to the production of electricity in 2014 (Wirth, 2015). The total electrical load in Germany presently varies from 30 to 80 GW during the day, regularly leading to an excess in electrical energy at noon, which could in principle be stored in form of chemical energy. Despite these high renewable energy capacities, renewable energies in Germany only contribute by 27% to the gross electric power consumption including biomass (8.8%) and water power (3.4%). Other energy sources still are nuclear energy (15.9%) and fossil fuels such as brown coal (25.6%), black coal (18%) and natural gas (9.6%). These numbers document the still massive dependency of industrial countries from fossil fuels. Nevertheless, renewable energies in Germany contribute by 11% to the primary energy consumption

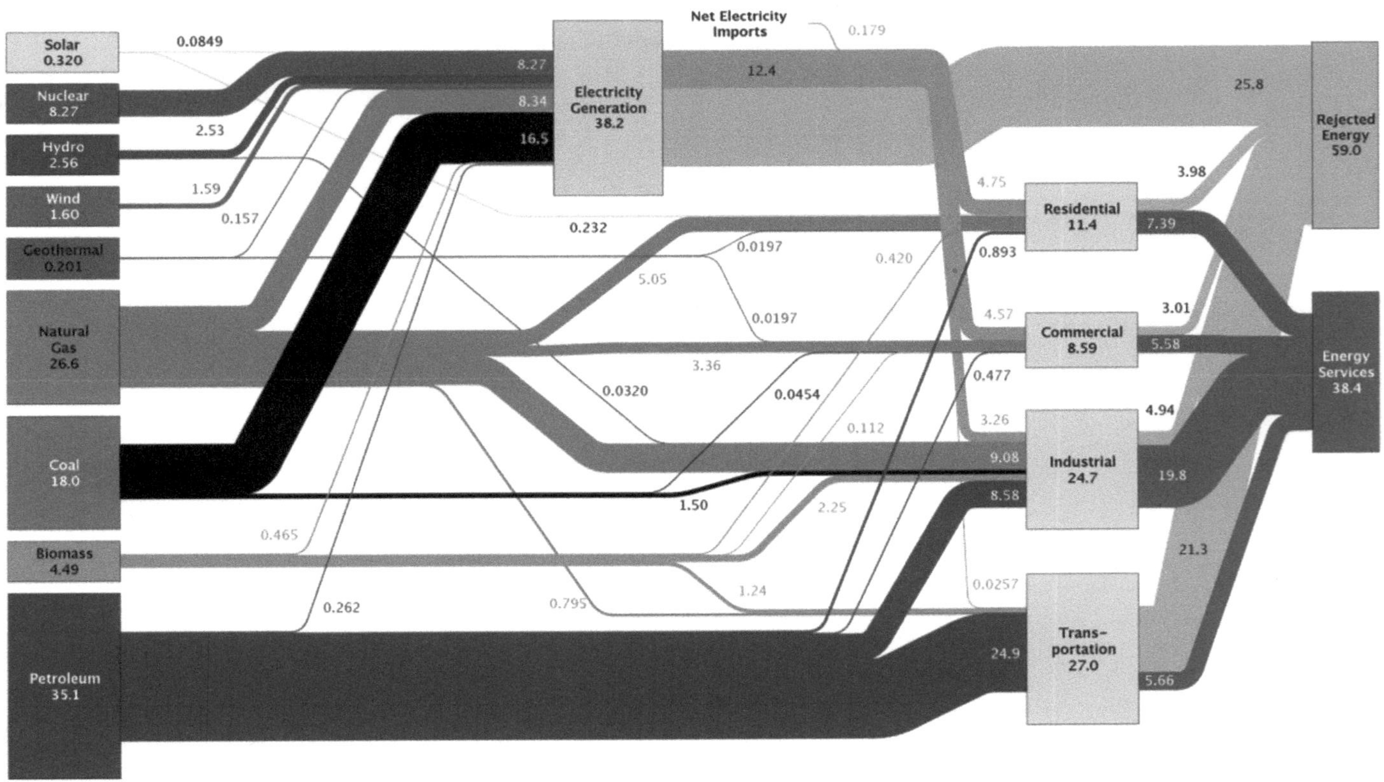

Figure 2 Energy flow diagnostics of estimated US energy use in 2013 in QUADS (1 QUAD = 293.1 TWh). © *Lawrence Livermore National Laboratory and Department of Energy.*

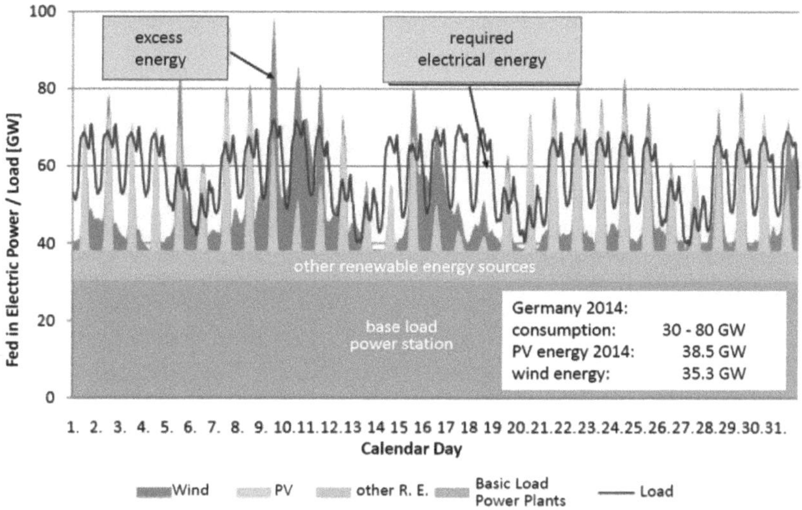

Figure 3 Injected electrical power in the grid and typical load behaviour. *Adapted from Brendli, J. (2011).* Storage of wind power and solar power within the existing gas grids as a basic concept in the change of energy supply towards renewable energy — Demand and comprehension of accumulating energy *(Bachelor thesis, Hochschule München/Thüga AG).*

(Bundesministerium für Wirtschaft und Energie, 2015). To totally cover it by renewable energies, its capacity has therefore to be increased at least by one order of magnitude.

The high capacities in wind and solar energy are regularly leading to an excess energy in electricity which has to be exported to neighbouring countries. In a future scenario, this energy could also be directly or indirectly converted into chemical energy, such as hydrogen, which will be stored chemically (eg, in hydrides) and physically (gas tanks), respectively. This situation is illustrated in Fig. 3 where peaks in wind and solar energy surpass the electrical energy need of a sunny or windy day (Brendli, 2011). In this diagram, other renewable energies illustrated by the green stripe, such as biomass and water power, do not show day and night cycles, but are of course also dependent from the seasons. Base load power stations are again nuclear and fossil power plants. The contribution of electricity produced by them is smaller than present, assuming a further increase of renewable energies in Germany.

To convert wind energy into hydrogen, advanced electrolysers must be used. Of special interest are polymer electrolyte membrane (PEM) electrolysers, which are commercially available producing presently 65 kg H_2 per day (Proton OnSite) as recently reported by Ayers, Danilovic, and Anderson

(2014). Lewinski and Luope (2014) discussed current densities >10 A/cm^2 in these kind of devices. Since they should be able to operate at high current densities and under variable load conditions when coupled with intermittent power sources such as wind generators and solar cells, PEM electrolysers presently are the best systems to meet these requirements (Rau et al., 2014). However, the high costs of its components (noble metal catalysts, NAFION membranes, Ti materials) have so far impeded large-scale commercial applications. Therefore, also other options have to be taken into account and to be developed. Alternatives could be integrated devices which can switch from an electricity-producing to a hydrogen-generating mode, the characteristics of which will be discussed in the next chapters.

2. ARTIFICIAL PHOTOSYNTHESIS

Plants, green algae and cyanobacteria are able to absorb sunlight and to convert it into chemical energy by producing hydrogen, starch and sugars in the process of photosynthesis. Educts needed are water and carbon dioxide and products are oxygen and a related fuel. Approaches to mimic the natural processes happening in photosystem I (PSI) and II (PSII) integrated in the thylakoid membrane of the chloroplasts by human-made systems is addressed as artificial photosynthesis (Reece et al., 2011).

Worldwide research activities in this field can be divided into five groups:

1. fuel generation using gene modified microorganisms,
2. fuel generation by hybrid systems immobilizing enzymes on electrodes,
3. fuel generation using metal organic molecules,
4. fuel generation using photochemically/catalytically active surfaces of semiconductor particles and
5. photoelectrochemical/photoelectrocatalytic devices using semiconductors equipped with catalysts.

Ciamician was convinced that photochemistry should be used 'to imitate the assimilating processes of plants'. In 1912 he wrote 'that we can modify to a certain extent the chemical processes that take place during the life of the plants' (Ciamician, 1912). Such an approach is illustrated in Fig. 4.

In the upper left corner of this figure, the principle features of a water splitting system are depicted. It consists of an absorber P (eg, a pigment or chromophore connected to an antenna system to capture photons as known from PSI and PSII), an electron donor D and an electron acceptor A. Both

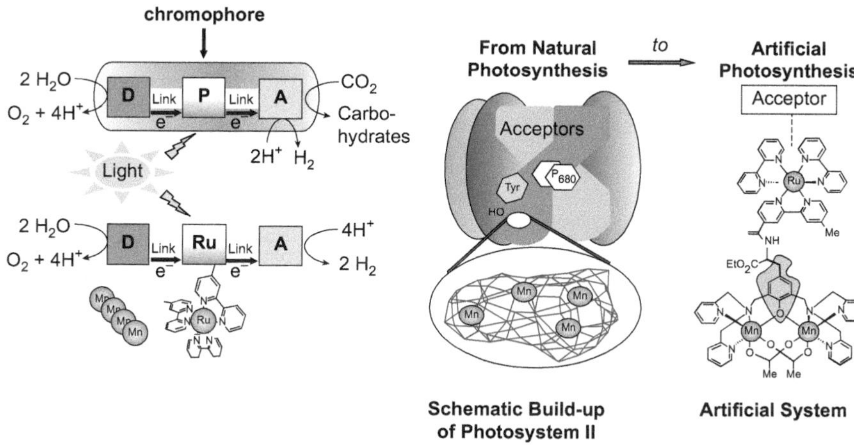

Figure 4 Schemes of artificial leaf-type systems for light-induced water splitting using molecules and transition metal complexes. *Reproduced by permission of Stenbjörn Styring, Uppsala University, Sweden.*

are linked to the absorber P by chemical bridges, which under illumination generate light-excited electron–hole pairs. After excitation the chromophore is able to uptake electrons released by the catalytic centre D (electron donor). The light-excited electrons migrate to the catalytic centre A (electron acceptor), where protons are reduced under formation of hydrogen or where protons, electrons and CO_2 can react under formation of a carbohydrate. Instead of the pigments in the protein matrix of PSI and PSII, a simplified artificial system can be built replacing P by, eg, a transition metal organic complex linked by a bridge to a metal organic Mn complex as known from PS II (Mn_4Ca complex, as shown in the middle of Fig. 4) acting as oxygen evolving catalyst (OEC). Instead of a tetramer in PSII a dimer Mn complex is shown in Fig. 4 at the right side. As acceptor, a metal organic iron complex can be used again linked to the photoactive absorber P (Styring, 2015). Not shown in this simplified picture is the migration of protons from D to A, which is in the case of the thylakoid membrane a complex process. In artificial systems the distance from D to A should spatially be as short as possible because the diffusion velocity of protons in aqueous electrolytes is essentially smaller than the transport of electrons.

Instead of a chromophore, also a thin semiconducting layer or an array of these layers can be used as light absorber generating excited electron–hole pairs of a potential large enough to split water into hydrogen and oxygen (Fig. 5). To overcome kinetic barriers, front and back contacts of the absorber layer have to be decorated by suited catalysts. Reece et al.

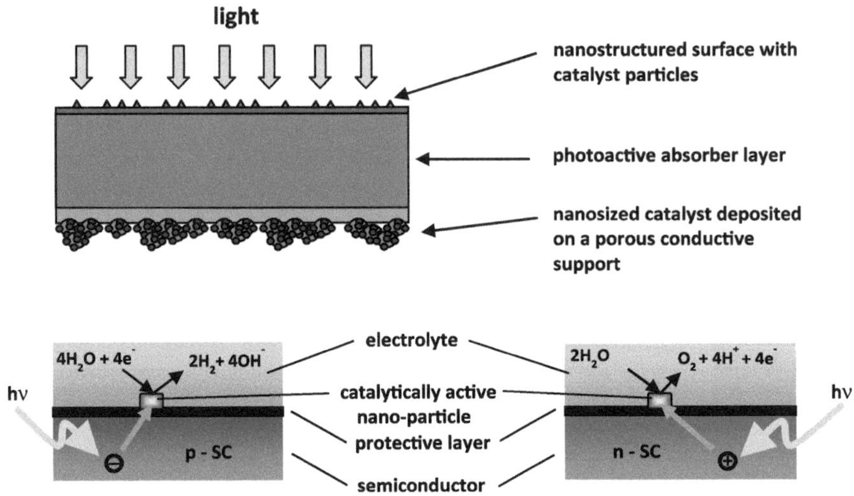

Figure 5 Artificial leaf-type structures: a light-absorbing photoactive layer deposited on a conductive and thin substrate is immersed in an electrolyte. To protect the water splitting unit against corrosion a transparent and conductive layer has to be deposited on top of the absorber layer which is equipped with catalyst particles, eg, the hydrogen evolution reaction (HER). Since four electrons have to be transferred to produce one oxygen molecule in a light-driven water splitting process, a highly porous and conductive substrate decorated by an OEC is advantageous. Since in the upper picture, H^+ ions in an acid or OH^- ions in a basic electrolyte have to migrate from the cathode to the anode side and vice versa lowering the efficiency of the device, arrays are of interest where cathodes and anodes are oriented ion parallel (lower part of the figure) to shorten the diffusion lengths of these ions in the electrolyte.

(2011) have addressed such a device, where all these components are deposited on a thin sheet of a metal foil (steel), an artificial leaf. Compared to other devices where (photo)anode and (photo)cathode are wired, such devices are called hybrid photoelectrodes.

Nature uses, in photosynthesis transition, metal complexes attached to the photosystems to evolve oxygen and hydrogen from water. In these processes the catalytic centres are separated from the light capturing and absorbing cofactors in PSI and II of the thylakoid membrane. An inorganic analogue is shown in Fig. 5. On a photoactive semiconductor layer on both sides, catalysts are deposited.

3. PHOTOELECTRODES AND DEVICE STRUCTURES

Which electrodes and which device structures are most promising realizing artificial leaf structures? Since in the last decades different types of solar cells have been developed for photovoltaic applications, one first strategy is to modify these cells using front and back contacts as electrodes provided with hydrogen and OECs, which have to be stable under acid, neutral or basic conditions without losing efficiency in the process of light-driven water splitting.

In the following, two strategies will be discussed:

- solar cell structures in superstrate geometry, ie, a geometry of the PV layer sequence where back and front contacts of the cell are oriented in the same plane or parallel to each other, not influencing the light absorption process in the photoactive part;
- solar cell and electrode structures, respectively, where the protecting as well as the catalyst layers are so thin that no light absorption influences the excitation of electron—hole pairs in the stacked semiconductor array.

In both cases it has to be kept in mind that the processes oflayer deposition on top of the catalysts have to be performed at temperatures not leading to any degradation of the semiconductor/back contact/catalyst junctions.

Fig. 6 is schematically showing such devices. Stellmach et al. (2013), Bogdanoff et al. (2015) as well as Urbain et al. (2015) used superstrate configurations modified by electrocatalysts deposited at back and front contacts. As mentioned before, the solar cell is illuminated via the glass substrate while the back and front contacts are arrayed laterally or in parallel. Therefore, neither shadowing effects nor light scattering by evolved gas bubbles can detrimentally affect the solar-to-hydrogen efficiency of this artificial leaf-type structure. By modifying the contact layers of the cell, its chemical stability in the electrolyte can significantly be improved (Bogdanoff et al., 2015).

The photocurrent densities of a-Si/μc-Si solar cells are typically restricted to photocurrent densities of 10 mA/cm^2 limiting the solar-to-hydrogen efficiency (η_{SHE}) of the hybrid electrolyser to about 10% (Bogdanoff et al., 2015; Urbain et al., 2015):

$$\eta_{SHE} = I_{sc} * U/P_{light} \tag{1}$$

whereas I_{sc} is representing the short circuit current density in mA/cm^2, $U = 1.23$ V the Gibbs free energy ΔG of the water splitting reaction at 25°C and P_{light} the intensity of the incident solar light of 100 mW/cm^2.

To remarkably increase η_{SHE}, solar cells and solar cell structures have to exhibit essentially higher photocurrent densities at a photovoltage of $\geq$1.6 V. A first demonstration of such a device was described by Khaselev and Turner (1998) and later by Licht et al. (2000). In both cases the authors used a highly efficient III-V based tandem and triple junction solar cell. In the

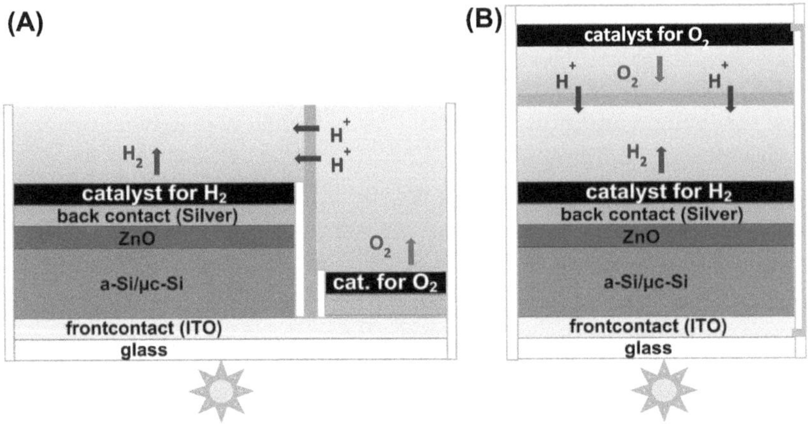

Figure 6 Hybrid electrolysers consisting (A) of a superstrate solar cell with a lateral array of cathode and anode deposited on a common transparent conductive substrate using a stack of amorphous/microcrystalline silicon layers (a-Si/µc-Si) in a multijunction configuration at the cathode side to obtain a photovoltage of at least 1.6 V and (B) using the same photocathode configuration, whereas the counter dark electrode equipped by an oxygen evolution catalyst is now oriented parallel to the back contact of the photocathode. In this configuration the anode is electrically connected to the front contact of the photocathode. In both cases, cathode and anode compartments are separated by an ion exchange membrane which is, in the illustrated case, an acid electrolyte. Illumination occurs via the glass substrate. *Modified after Stellmach, D., Bogdanoff, P., Gabriel, O., Stannowski, B., Schlatmann, R., van de Krol, & Fiechter, S. (2013). Nanostructured* MoS_2 *particles as a novel hydrogen evolving catalyst integrated in a PV-hybrid electrolyzer. In A. Méndez-Vilas (Vol. Ed.).* Materials and processes for energy: Communicating current research and technological developments *(pp. 880–886). FORMATEX.*

case of Turner's approach the photoelectrode was immersed in the electrolyte. In Licht's device the solar cell was kept outside the electrolyte and the Pt and RuO_2 electrode, respectively, were wired to back and front contacts of the cell. Peharz, Dimroth, and Wittstadt (2007) connected such a cell to a PEM electrolyser demonstrating an efficiency of at least 14%. May, Lewerenz, Lackner, Dimroth, and Hannappel (2015) published an integrated water splitting device as illustrated in Fig. 7 demonstrating an efficiency of 14%, the highest so far reported value for a device used as artificial leaf.

At present, the manufacture of III-V-based solar cells, which consists of epitaxial layers grown on a Ge single crystal substrate or another type of a single crystal substrate, are costly and therefore presently not suited for a mass production. For that reason, alternative materials and device structures have to be developed. One promising route is to further optimize semiconductive oxides and oxynitrides as light absorbers. Most of the novel metal oxides and

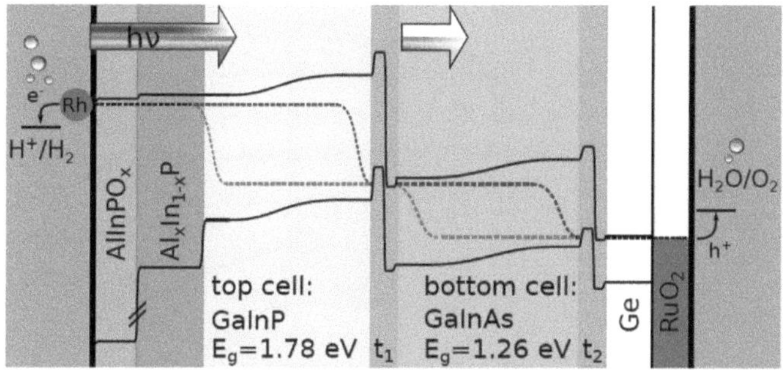

Figure 7 Energy diagram of a water splitting device consisting of an n-p GaInP solar cell connected by a tunnel junction to an n-p GaInAs structure. The ohmic back contact is wired to a sputtered RuO_2 counter-electrode acting as oxygen evolving unit. As hydrogen evolving catalyst rhodium nanoparticles were deposited photoelectrochemically on the front contact of the cell. *Dashed lines* indicate the Fermi and Quasi-Fermi levels of electrons and holes (May et al., 2015). *Permission by Creative Commons Attribution 4.0 International License.*

oxynitrides under investigation have complex crystal structures and compositions and are limited as semiconducting absorber materials by their chemical stability when brought in direct contact with an electrolyte used as a photochemically active electrode for water splitting. Most of these materials are characterized by small absorption coefficients, small electron and hole mobilities (<1 cm^2/V), small charge carrier lifetimes (ps to fs range) and short diffusion lengths for electrons and holes, which so far limits their applicability as electrode materials in water splitting devices. To overcome these drawbacks nano- and microstructuring of the electrode surfaces, texture and doping of the polycrystalline films are tools to significantly improve the optoelectronic properties. In the last decade, many groups concentrated on an improvement of the current densities of the absorber material neglecting the task how to also improve the photopotential of the developed photoelectrodes. For that reason, all these electrodes need a bias voltage delivered in most cases by a multijunction silicon solar cell. The deposition of appropriate electrocatalysts influences the electrode/electrolyte interface but can also passivate states at the electrode/catalyst interface but can also passivate states at the electrode/catalyst interface.

In principle a water splitting device could also consist of a p-type and an n-type electrode in a side by side configuration as displayed in Fig. 8. Both absorbers are protected by a transparent and conductive oxide layer protecting the absorbing oxide/oxynitride layer against corrosion as well as by a thin catalysts layer lowering the overvoltages in the processes of hydrogen

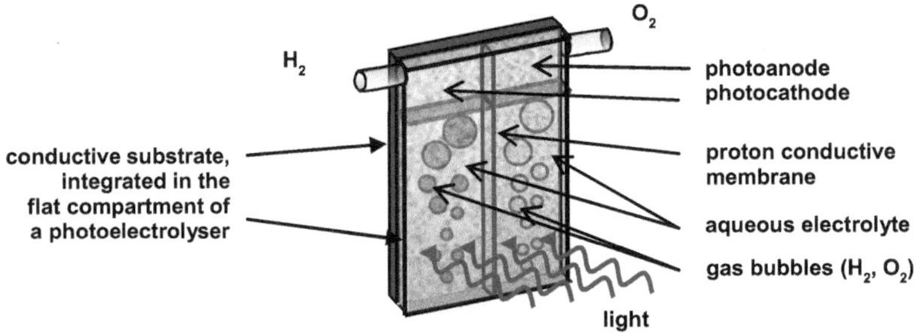

Figure 8 Schematic drawing of a water splitting device consisting of a side-by-side configuration of photocathode and photoanode.

and oxygen evolution, respectively. In all cases abundant, cheap and nontoxic elements are needed in the light-driven water splitting devices.

As p-type semiconducting oxides Cu_2O, $CuFeO_2$ and $CuAlO_2$, but also some ferrites or perovskites can be used. Fig. 5 is schematically showing a hydrogen evolving electrode. In the case of p-type Cu_2O a p-n junction is made using ZnO as emitter layer. Since ZnO has a very limited stability in aqueous electrolytes, a protective thin film of a highly doped TiO_2 layer was deposited on top which is addressed as buffer layer in Fig. 5. To lower overvoltages, platinum nanoparticles or as a noble metal free catalyst MoS_2 particles were deposited on top (Morales-Guio, Tilley, Vrubel, Grätzel, & Hu, 2014; Paracchino, Laporte, Sivula, Grätzel, & Thimsen, 2011). As a counter electrode, oxides such as Fe_2O_3 (haematite) or $BiVO_4$ have been developed (Fatwa et al., 2013; Kim et al., 2013) as absorber materials. Their electrode surfaces have been protected by the deposition of an OEC, such as a amorphous CoO_x layer. Since the optical absorption coefficients as well as the optoelectronic properties (electron and hole mobilities, charge carrier lifetimes etc.) of these materials are limited compared to silicon and III-V semiconductors, nanostructured electrodes have been considered as an approach to overcome these drawbacks (Pihosh et al., 2015, Fig. 9). It is expected that metal oxides and metal oxynitrides exhibit a higher chemical stability against corrosion — when immersed in an electrolyte in a wide pH range compared to diamond-based semiconductor materials. On the other hand, the conduction mechanism of these compounds is different from that of silicon and other semiconductors used for solar cells because of small conduction bandwidths. While excited electron—hole pairs in silicon can be considered as independently migrating in the volume of the semiconductor, charge transport in oxides has to be discussed in terms of a hopping mechanism whereas distortions in the crystal lattice are occurring during the migration of the charges. Under these conditions the migration of charges

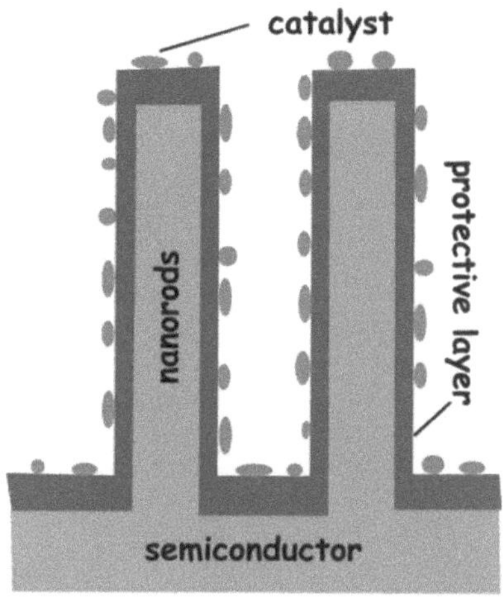

Figure 9 Scheme of a nano/microstructured electrode with a protective layer on top decorated by catalytically active nanoparticles. These types of structures can improve light absorption and electrode efficiency in the process of light-induced water splitting.

in the lattice is dependent from the charge transport in the neighbourhood, and the transport of charge carriers is therefore no longer independent of each other. However, an advantage of using complex metal oxides/oxynitrides is to tailor the position of conduction and valence band related to the redox potentials of water splitting, namely H^+/H_2 and O_2/H_2O.

Figure 9 schematically shows a substrate (dark yellow) with nanorods grown on top, the length of which can amount to several micrometres. Doping of the semiconductor nanorods has to be chosen related to the mobility of electrons and holes in the material (Kim et al., 2013). If the mobility of the holes in the material is small the rods should be doped n-type to act as photoanode, in case the mobility of the electrons is low an opposite doping is advantageous to use this structure as photocathode. Under these conditions, one type of charge of low mobility has to only migrate a short distance from the bulk of the nanorod to the surface first passing the protective layer (blue) before injected into the catalyst nanoparticles (shown as pale red particles in Fig. 9) to finally react at the catalyst/electrolyte interface with water under formation of oxygen and hydrogen, respectively. The protective layer can also be composed of a second semiconductor characterized by an energetic position of valence and conduction band energetically higher or lower in comparison to the light-absorbing nanorod semiconductor structure to improve charge carrier transport to the surface. In Fig. 10, such a structure is illustrated (Pihosh et al., 2015). A thin layer of $BiVO_4$ was grown

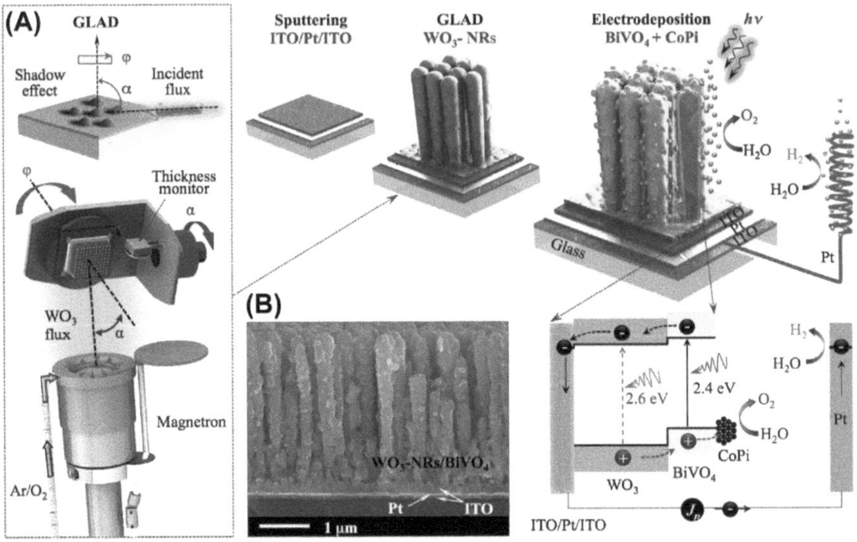

Figure 10 Illustration of a core-shell $WO_3/BiVO_4$ structure consisting WO_3 nanorods fabricated by glancing angle deposition (GLAD) (A) using magnetron sputtering and covered by an electrodeposited $BiVO_4$ layer (Pihosh et al., 2015). Finally an amorphous CoO_x (CoPi) layer was deposited on top (upper right part) following a recipe described by Reece et al. (2011). (B) a scanning electron micrograph of the cross section of the oxygen evolving electrode is shown: the WO_3 nanorods (NRs) are covered by a thin film of $BiVO_4$ and by nanoparticles of the OER catalyst CoPi. In the lower right corner the energy band diagram of the oxygen evolving nanorod structure is shown. *Permission by Creative Commons Attribution 4.0 International License.*

electrochemically on top of conductive WO_3 nanorods. To lower the overvoltage in the process of water oxidation, an amorphous CoO_x film was deposited on top of the $BiVO_4$ layer. Since the energetic positions of valence and conduction band of $BiVO_4$ are located slightly higher than those of WO_3, an improved transport of holes to the CoO_x/electrode interface is thought to be responsible for the improved current densities of the photoelectrode. With this heterojunction, Pihosh et al. (2015) were able to demonstrate photocurrents of up to 6.72 mA/cm^2 at 1.23 V_{RHE}, the electrochemical potential related to the Gibbs free energy of water splitting. The obtained current density corresponds to ~90% of the theoretically possible value for $BiVO_4$. Since the photovoltage of this electrode is by far too low to split water, the photoanode was wired with a GaAs/InGaAsP photovoltaic cell to demonstrate a solar to hydrogen efficiency of 8.15%.

The electrode geometry as shown in Fig. 10 has been originally proposed by N. Lewis et al. (Shaner et al., 2013). They emphasized that a nearly regular array of nanowires acts as light trap and is in addition decreasing the semiconductor thickness, normally required in a nonstructured absorber layer, by a factor of $4n^2$, where n is the refraction index. Since photogenerated minority charge carriers only have a short distance to travel to the reactive interface, the requirements on the purity of the semiconductor material are also not as high as for a bulk absorber layer.

However, a high number of recombination centres on the increased electrode surface of the nanostructured electrode has to be passivated growing — for example — a thin conformal, but highly conductive and transparent layer on top of the structured electrode surface which can be achieved by atomic layer deposition (ALD) technique. The expected advantages using nanostructured devices in terms of an increased light trapping as discussed earlier have so far only be achieved in the case of the $WO_3/BiVO_4/CoO_x$ and α-Fe_2O_3/CoO_x nanowire electrodes (Kim et al., 2013; Pihosh et al., 2015). To further improve this promising type of photoelectrodes, novel semiconducting materials have to be developed possessing a band gap <2 eV. They should also consist of cheap, abundant and nontoxic elements showing a solar-to-hydrogen (STH) efficiency >10%, a precondition to generate hydrogen at a price of ≤3 €/kg H_2. Unfortunately, the band gaps of WO_3 ($E_g = 2.6$ eV) and $BiVO_4$ ($E_g = 2.4$ eV) are too large to obtain these efficiencies, but α-Fe_2O_3 with a band gap of $E_g \sim 2.2$ eV could in principle achieve such an efficiency.

4. APPLICATION OF ENERGY-CONVERTING DEVICES FOR CO_2 REDUCTION

Artificial leafs can be used to produce hydrogen by water splitting, but in principle also hydrocarbons can be accomplished by a light-driven reaction from CO_2 and protons at the cathode. As known, these products are feedstocks for fertilizers, pharmaceuticals, plastics and synthetic fuels. Hydrogen can be directly used as a fuel operating electrical vehicles equipped with a fuel cell converting chemical energy into electricity.

The steadily increasing CO_2 concentration in our atmosphere is thought to be responsible for the ongoing dramatic climate change on earth. To stop its further increase, CO_2 from the atmosphere has to be absorbed and reduced by, eg, production of fuels such as HCOOH, CH_4 or CH_3OH using renewable energy sources in a closed cycle (Fig. 11). In nature, photosynthesis is able to extract CO_2 from the atmosphere and to produce under reaction with water hydrocarbons by converting sunlight into chemical

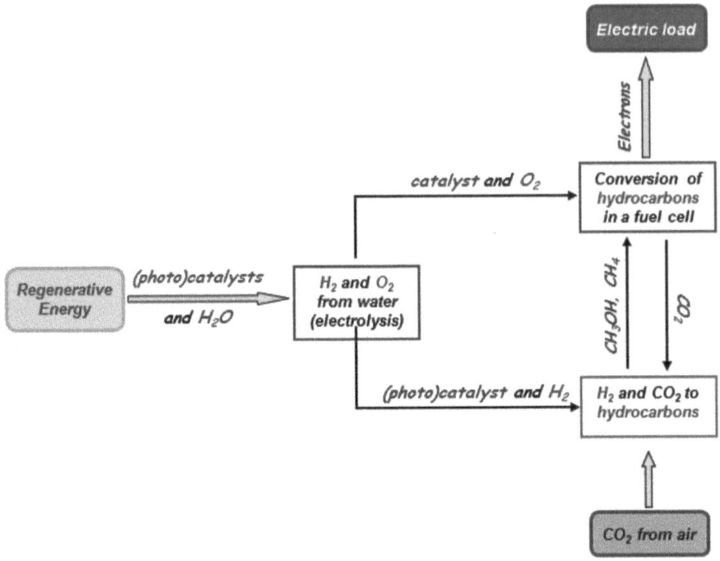

Figure 11 Closed energy cycle for CO_2 reduction: Photoelectrochemically generated hydrogen is used to produce hydrocarbons directly or in a thermally activated dark process after light-driven hydrogen generation.

energy in form of starch and sugars. Besides this process, nature is also able to store CO_2 from the atmosphere in the oceans in large quantities. However, natural storage systems have so far not been able to buffer the steadily increasing CO_2 concentration present in the atmosphere produced by mankind in the last 200 years.

Due to the complexity of the reaction of water with CO_2 under formation of fuels, the efficiency of the process, which even in nature only amounts to 1% in photosynthesis, related to starch and sugar production, has to be increased.

Doing it (photo)electrochemically the process can be split into a light-driven part and a process occurring in the dark as known from photosynthesis (water splitting at the thylakoid membrane and starch formation in the Calvin cycle). In the light-driven cycle of an artificial system, hydrogen can be produced using efficient photoelectrodes decorated by electrocatalysts, which will be converted to hydrocarbons in a subsequent dark process as known from Sabatier and Fischer–Tropsch process (methane and diesel production and reverse water gas shift reaction, respectively).

The development of efficient photocatalysts or photoelectrocatalysts for CO_2 reduction is one of the most challenging tasks in artificial photosynthesis.

Consideration of the thermodynamics of CO_2 reduction elucidates the complexity of the process. Eqs (2)—(6) indicate that the potential E^0 needed to electrochemically convert CO_2 into higher energetic products is small. However, the thermodynamic considerations of the reactions neglect that in a light-driven process, one photon can only lead to one electron transported per reaction step to finally form the products such as formic acid (Eq. (2)), carbon monoxide (Eq. (3)), formaldehyde (Eq. (4)), methanol (Eq. (5)) and methane (Eq. (6)). For that reasons, the voltages, i.e. the overvoltages, necessary to drive the processes are essentially higher than the thermodynamic potentials as summarized in Eqs (2)—(6). In contrast, hydrogen evolution (Eq. (7) for pH 7) at Pt catalysts only needs a small overvoltage.

$$CO_2 + 2H^+ + 2e^- \rightarrow HCOOH \quad E^0 = -0.61\ V \tag{2}$$

$$CO_2 + 2H^+ + 2e^- \rightarrow CO + H_2O \quad E^0 = -0.53\ V \tag{3}$$

$$CO_2 + 4H^+ + 4e^- \rightarrow CH_2O + H_2O \quad E^0 = -0.48\ V \tag{4}$$

$$CO_2 + 6H^+ + 6e^- \rightarrow CH_3OH + H_2O \quad E^0 = -0.38\ V \tag{5}$$

$$CO_2 + 8H^+ + 8e^- \rightarrow CH_4 + 2H_2O \quad E^0 = -0.24\ V \tag{6}$$

$$2H^+ + 2e^- \rightarrow H_2 \quad E^0 = -0.42\ V \tag{7}$$

The oxidation of water at the counter electrode is used as a source for electrons and protons, which are needed to reduce CO_2. This process is described by Eq. (8) and relates to a potential of 0.81 V at pH 7:

$$2H_2O \rightarrow O_2 + 4H^+ + 4e^- \quad E^0 = 0.81\ V \tag{8}$$

Due to the four electrons, which have to be transferred in this process, overvoltages in the range from 0.3 to 0.4 V are necessary to move an oxygen evolution at a current density of $<10\ mA/cm^2$.

4.1 Photocatalysts in Homogeneous Systems

Molecular and supramolecular metal organic complexes such as imides, amides, cyclam, phenanthrolines, dipyridines, porphyrins etc., with a transition metal (ruthenium, rhenium, cobalt and nickel) in the centre of the complexes were intensively investigated as catalysts. In homogeneous systems, some of these photocatalysts show high absorbance in the visible region, high quantum yields and high product selectivity. However, no metal complex which can use water as reductant according to Eq. (8) for CO_2 reduction has been reported so far (see Yui, Tamaki, Sekizawa, & Ishitani, 2011).

4.2 Heterogeneous Catalysts to Catalyse Hydrocarbons in the Dark

Very important reactions using metal oxide supported Cu, Cu_2O and Ni catalysts in the formation of methanol and methane from CO_2 and H_2 include the following:

$$CO_2 + 3H_2 \rightarrow CH_3OH + H_2O \quad \Delta H^{298.15K} = -87.785\ kJ/mole$$

$$CO_2 + 4H_2 \rightarrow CH_4 + 2H_2O \quad \Delta H^{298.15K} = -165.063\ kJ/mole$$

Alkene formation from CO_2 and H_2 is catalysed by Fe and Co catalysts in the Fischer–Tropsch process. In this context the water gas shift reaction,

$$CO_2 + H_2 \rightarrow CO + H_2O \quad \Delta H^{298.15\ K} = +41.137\ kJ/mole$$

plays an important role followed by CO dissociation via the Boudouard equilibrium and subsequent hydrogenation.

4.3 Heterogeneous Systems Using Photoactive Semiconducting Electrode Surfaces

Monoclinic $BiVO_4$ can photocatalyse CO_2 reduction to C_2H_5OH in aqueous solution without a sacrificial electron donor (Liu et al., 2009). The oxide electrode can act as visible light responsive photocatalyst, where the photocatalytic activity is improved under intense light irradiation. CO_2 reduction was also observed on Cu_2O alloyed TiO_2 electrodes. In 2011, Li et al. described methanol formation on Cu_2O/SiC nanoparticles under visible light irradiation.

In spite of the extensive literature on carbon dioxide reduction on copper electrodes, the detailed mechanism of hydrocarbon formation is still unclear. Hori (2012) showed that the extent of methane and ethylene formation sensitively depends on the surface orientation of the copper electrode.

Possible research for the development of a cost-effective electrocatalyst for CO_2 reduction points at copper and Cu_2O as promising catalyst materials. In the technical production of methanol from carbon dioxide and hydrogen, copper nanoparticles supported on ZnO/Al_2O_3 were used successfully. However, this process requires the utilization of

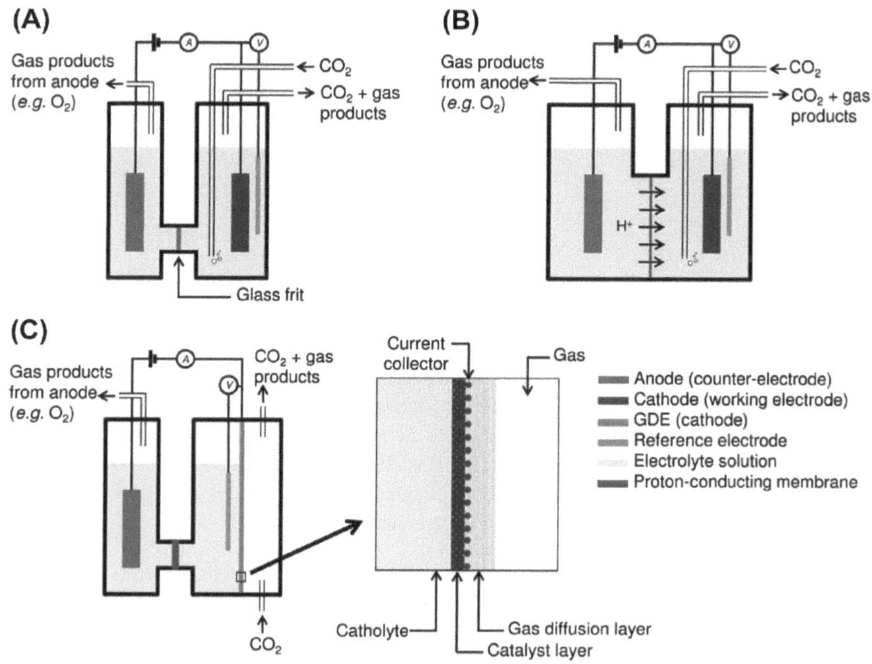

Figure 12 Architecture of devices which in principle allow the photoelectrochemical reduction of CO_2 to hydrocarbons using as power source an external solar cell or a photoelectrode with integrated catalysts (Kondratenko, Mul, Baltrusaitis, Larrazábal, & Pérez-Ramírez, 2013). (A) is showing a two-compartment cell, (B) a cell with electrodes separated by an H^+-conducting membrane, and (C) a cell using a gas diffusion electrode.

pure CO_2 under high pressure and at temperatures >150°C (Behrens et al., 2012).

Therefore, a starting point of research could be the deposition of nanoscaled Cu and Cu_2O particles onto selected photoactive semiconducting electrode surfaces such as SiC or $BiVO_4$ to test their behaviour in photoelectrochemical devices as shown in the drawings of Fig. 12. As electrolytes, aqueous, but also organic liquids can be used having the advantage to solve a higher concentration of CO_2 than aqueous electrolytes.

4.4 Solar Cells Combined With an Electrochemical Unit

Recently Schreier et al (Marcel Schreier et al., 2015) demonstrated a solar-to-CO efficiency exceeding 6.5% employing series-connected perovskite solar cells connected to an electrochemical device with IrO_2 (OEC) and Au catalytically active electrodes (see Fig. 13).

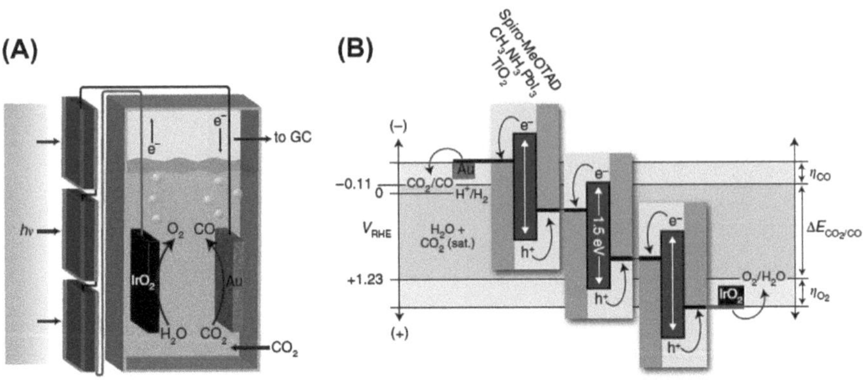

Figure 13 Sunlight-driven CO_2 reduction device using three in series-connected $CH_3NH_3PbI_3$ perovskite-type solar cells. (A) Schematic of the reduction device; (B) Band diagram of the solar cells wired to the Au − IrO_2 electrodes (Marcel Schreier et al., 2015). Obviously, an overvoltage of about 3 V is here necessary to convert CO_2 into CO. *Permission by Creative Commons Attribution 4.0 International License.*

5. PRESENT RESEARCH TOPICS

Presently worldwide, the following research topics are under consideration:

- development of photocatalysts and (photo)electrocatalysts, which have to be stable in a wide pH range under hydrogen and oxygen evolution conditions;
- development of photoactive semiconductor particles, the crystal facets of which are covered by appropriate hydrogen evolving catalysts (HEC) and OEC;
- structured photoelectrodes used as electrochemical half cells which are protected by transparent and conductive buffer layers and thin catalyst films protecting the electrode against electrochemical corrosion;
- multijunction solar cells of high efficiencies wired with novel water splitting or CO_2 reducing electrodes and
- multijunction solar cell devices in superstrate configuration with corrosion stable front and back contact.

McCrory et al. (McCrory, Jung, Peters, & Jaramillo, 2013; McCrory et al., 2015) evaluated noble metal free transition metal alloys and transition metal oxides as well as CoP_x and CoB_x as electrocatalysts with respect to their stability as OEC and HEC under acid and alkaline conditions, respectively. He checked their electrochemical stability at a current density of 10 mA/cm^2 under illumination with one sun for 2 h. As best materials for

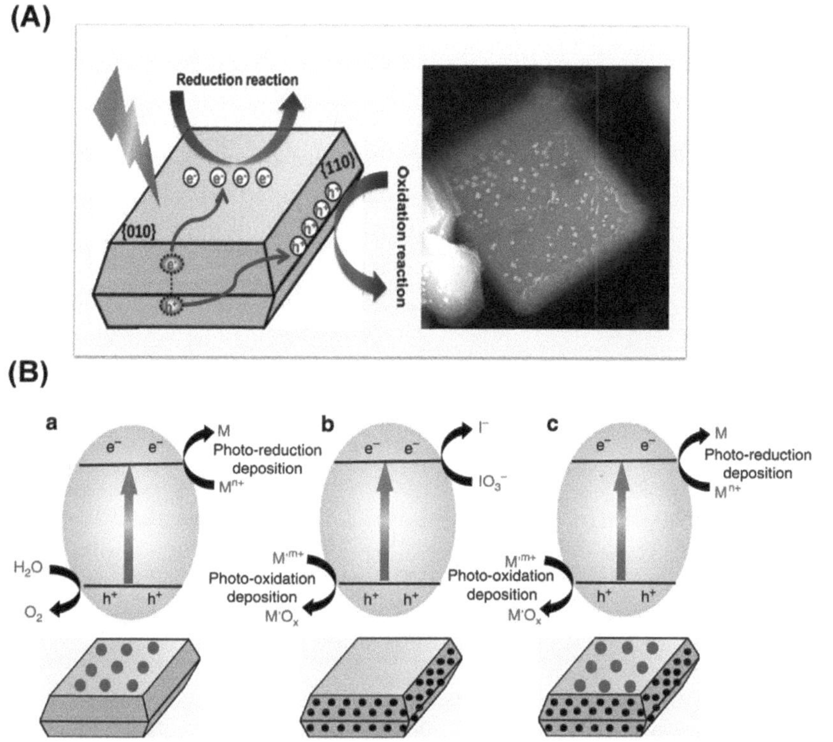

Figure 14 A monoclinic $BiVO_4$ particle (A) which shows under illumination a positive charging of the {110} and a negatively one of the {010} facets. This behaviour can be used to photoelectrochemically deposit MnO_x on {110} and Pt nanoparticles on {010} facets of monoclinic $BiVO_4$ crystallites (B) (Li et al., 2013). *Permission by CC BY 4.0.*

the oxygen evolution reaction, he mentioned NiCo, CoP_x and $Fe_xNi_{1-x}O_y$ highest catalytic activities, which are however stable under alkaline conditions. The overvoltages ranged from 0.3 to 0.4 V. It has explicitly to be pointed out that good earth-abundant catalysts for the OER in acidic conditions are still missing, an important open question and challenge for the field. Towards HER, best results were obtained using metal alloys belonging to the systems Co-Mo, Ni-Mo and Ni-Mo-Co which were stable in acidic as well as in alkaline solutions showing an overvoltage of ~0.1 V.

Studying the deposition of catalysts on top of a photoactive oxide, Li et al. (2013) recently described the behaviour of monoclinic $BiVO_4$, tested as photoanode and as photoactive powder using facetted crystallites of 1–2 μm edge length. Under an illumination with a xenon lamp, MnO_x nanoparticles were deposited on {110} facets and Pt on {010} facets (Fig. 14).

Best conditions for water oxidation under irradiation were obtained using a Pt{010}/MnO_x{110}/$BiVO_4$ slurry suspended in a 0.02 M $NaIO_3$ aqueous solution. Deposition of catalysts by a uniform impregnation of all facets resulted in a very low catalytic activity towards oxygen evolution. Investigating photoelectrodes, best results were obtained after deposition of MnO_x on {110} facets while IrO_2 deposited on {010} facets led to a quenching of the oxygen evolution behaviour. This result is unexpected because IrO_2 is belonging to the best performing oxygen evolution catalysts. The previously described behaviour demonstrates that catalysts performing very effectively in the dark, show a totally different behaviour in a photoelectrochemical device when deposited on the wrong facets of the crystallites of the photoactive absorber material.

To use multijunction, a-Si/μc-Si solar cells in superstrate configuration in artificial leaf-type structures (Fig. 6), the standard ZnO/Ag layer (Bogdanoff et al., 2015; Stellmach et al., 2013; Urbain et al., 2015) has to be replaced by a modified back contact to increase long-term stability. Replacing it by a titanium layer, the efficiency of the PV cell drops by 60% mainly caused by the lower specular reflectivity of Ti compared to Ag. This modification stabilizes the device used as an artificial leaf for at least 20 h using Pt and RuO_2 nanoparticles as electrocatalysts deposited on front and back contacts (Bogdanoff et al., 2015; Stellmach et al., 2013). To lower the costs of the device, platinum could be replaced by MoS_2 as hydrogen evolving catalyst. For that reason, MoS_2 particles were prepared by wet chemical deposition technique. Best results were found employing multiwalled carbon nanotube supported MoS_2 nanosheets, which have the advantage of a higher corrosion stability towards the HER. Deposited as a blend with PEDOT:PSS on the backside of the solar cell, an STH efficiency of 3.7% was yielded. RuO_2 can be replaced by MnO_x and CoO_x, respectively (Stellmach, 2014).

6. TECHNICAL AND SCIENTIFIC CHALLENGES

Ager, Shaner, Walczak, Sharp, and Ardo (2015) reviewed laboratory demonstrations of spontaneous photoelectrochemical solar water splitting cells since 1976. In most cases, a photoactive semiconducting oxide electrode was wired either with a PV solar cell (Si- and/or III-V-based cells) and/or with a dark electrode. In 2000, Licht et al. presented a lab device using an III-V/Si monolithic PV cell wired with the electrodes of an electrolyser. In this work, a solar-to-hydrogen efficiency (SHE) of $\eta = 18.3\%$ and a photocurrent stable for 14 h were demonstrated. Two years before, Khaselev and Turner had realized an III-V-based photoelectrode with an integrated Pt-catalyst immersed to an electrolyte with a wired Pt counter electrode of $\eta = 12.4\%$, which was stable for 20 h (Khaselev & Turner, 1998). The efficiency of this cell could be exceeded by May et al. (2015) ($\eta = 14\%$).

To lower the high costs of III-V-based solar cells as water splitting electrodes, dye sensitized oxide electrodes using WO_3 and Fe_2O_3 as photoactive absorber layers were proposed and tested by Brillet et al. (2012) ($\eta = 3.1\%$). Besides this class of materials, halide perovskites ($CH_3NH_3PbI_3$) have been tested wired to NiFe-based electrocatalyst electrodes (Luo et al., 2014). An solar to hydrogen efficiency of 12.3% was obtained. In both cases, long-term stabilities are an issue.

So far not mentioned are thin film solar cells based on chalcopyrite which were successfully wired with dark electrodes of an electrolyser (Jacobsson, Fjallstrom, Sahlberg, Edoff, & Edvinsson, 2013).

Aim of all research efforts is to simplify the architecture of the devices and to reduce the number of components of the light-driven electrolysers targeting on an artificial leaf-type structure consisting of cheap, abundant and nontoxic components. The developed electrodes should be long time stable in contact with an electrolyte. The evolved gases have to be separated by an H^+ and OH^- conductive membrane, respectively. Instead of noble metals, alternative catalysts which can be deposited on the photoelectrode surfaces have to be developed.

Fig. 15 shows an example how such a light-driven electrolyser could be constructed. Here, the device consists of a triple junction thin film Si/Ge-based PV solar cell preventing the direct contact of the solar cell with the electrolyte. As substrate for the cell, a steel foil is used. As noble metal-free catalysts, a CoMo alloy for hydrogen evolution and a $Fe_xNi_{1-x}O$ for oxygen evolution are projected. As gas separator a thin membrane material has been chosen, mechanically stabilized by a plastic mesh.

To obtain a photovoltage of at least 1.6 V at a current density of 10 mA/cm^2, a tandem or a triple junction solar cell as photoelectrode has to be used (Bogdanoff et al., 2015; Urbain et al., 2015). In addition, the catalysts applied in both chambers of the electrolyser should be stable at the same pH value, ie, under acidic, neutral or alkaline conditions. Presently, OECs as alternative to RuO_2 and IrO_2 are only stable under neutral or alkaline conditions. Most prominent for this purpose are metal alloys (McCrory et al., 2013) and $Fe_xNi_{1-x}O$ catalysts (McCrory et al., 2015). However, HECs such as MoS_x

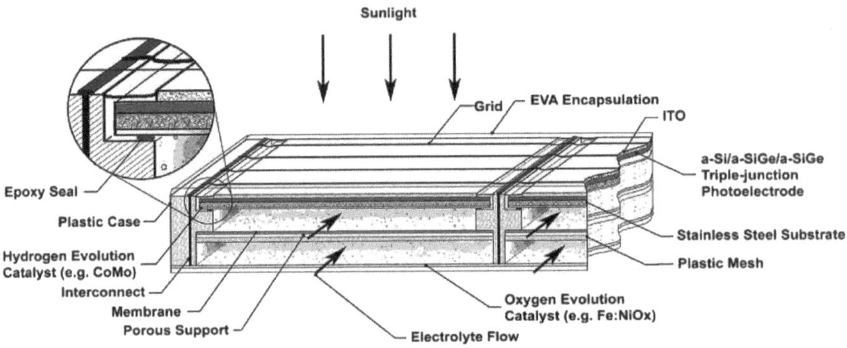

Figure 15 Array of flat and in-parallel oriented light-driven electrolyser units. The photovoltage of each water splitting device is generated by a triple junction a-Si/a-SiGe/a-SiGe thin film cell. Front contact of the PV cell is wired to the anode side (interconnect). The oxygen and hydrogen evolving compartments are separated by an H^+ or OH^- conductive membrane. *With permission from Midwest Optoelectronics, LLC and University of Toledo.*

demonstrate stability only under neutral or acidic conditions which could replace platinum (Stellmach, 2014; Stellmach et al., 2013).

A costly solution is the application of III-V-based solar cell structures wired to the electrodes of a PEM electrolyser as demonstrated by Peharz et al. (2007) but highest efficiencies were obtained so far with these devices. Nakamura et al. (2015) achieved the remarkable solar to hydrogen efficiency of 24.4% testing outdoor a concentrator photovoltaic module using InGaP/GaAs/Ge triple-junction cells. With these devices, current densities of 330 mA and photovoltages of 2.5 V have been accomplished.

6.1 Open Scientific Questions

In principle, a photoelectrode with a band gap ~2.2 eV should be able to split water without any bias voltage when the positions of valence and conduction band are energetically positioned below and above the redox potential for water splitting (Fiechter, Bogdanoff, Bak, & Nowotny, 2012; Kaiser, Jaegermann, Fiechter, & Lewerenz, 2011). Evaluating the energetic positions of valence and conduction band of several compound semiconductors with respect to the redox potentials H^+_{aq}/H_2 and O_2/H_2O, materials, such as $Ga_{1-x}In_xN$, C_3N_4 and SiC (see Fig. 16), should be able to unbiased split water under solar irradiation. In 1972, Fujishima and Honda demonstrated for the first time a water splitting photoelectrode using a single crystal of TiO_2 immersed in an electrolyte and irradiated with UV light (Fujishima &

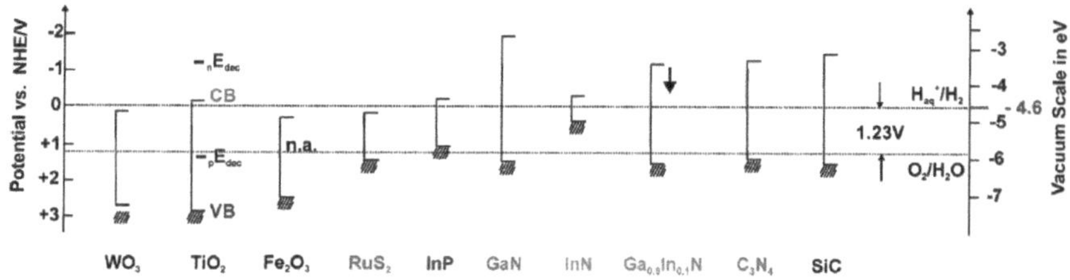

Figure 16 Valence and conduction band positions of different compound semiconductors with respect to the redox potentials H^+_{aq}/H_2 and O_2/H_2O at pH 7.

Honda, 1972). Although the band gap of TiO_2 (anatase) should be large enough, they only noticed water splitting when using two different pH values in the compartments of their photoelectrochemical cell.

Interestingly, the expected photovoltage following the equation

$$U_{oc} = E_g - 0.4\ eV \tag{9}$$

could, with the exception of perovskite-type $CH_3NH_3PbI_3$ (Kondratenko et al., 2013), so far not be verified in semiconductor materials of band gaps >1.3 eV. Two strategies to increase photoeffects are described in the literature:

- incorporation of a ferroelectric layer in the photoelectrode and
- passivation of surface states by depositing a thin transition metal oxide layer.

Photovoltaic effect of $BiVO_4$ semiconductor on the size of the photovoltage was investigated by incorporating an ultrathin $BiFeO_3$ ferroelectric layer in the electrode. Dong et al. (2013) could show that the ultrathin ferroelectric layer with strong self-polarization and high carrier density was able to enhance the photovoltaic effect and to manipulate the photovoltaic polarity of the semiconductors.

Yang et al. (2013) described the deterioration of a photoactive α-Fe_2O_3 semiconductor electrode after deposition of an ultrathin catalyst layer on top of the photoelectrode. They concluded that the surface treatment of the absorber layer was influencing the semiconductor/electrolyte interface. Depositing an ultrathin MnO_x by ALD on haematite, a substantial reduction in photovoltage was observed, the origin of which is explained by a Fermi-level pinning effect due to the catalyst.

Totally different behaved the photoelectrochemical performance of α-Fe_2O_3/CoO_x composite photoanodes (Zhong & Gamelin, 2010). The photoelectrocatalytic behaviour of this composite photoanode could be enhanced fivefold relative to that of a pure Fe_2O_3 electrode.

This limitation is generally discussed in terms of recombination centres in the bulk and on the surface of the photoactive electrode in contact with the electrolyte at the solid liquid junction (SLJ). An improvement could be expected depositing a thin layer of a catalyst on top of the semiconductor to passivate surface states on one hand and to lower the overpotentials occurring at the SLJ in the process of water splitting on the other hand. Especially, CoO_x and MnO_x particles deposited on semiconducting oxides are under inspection to act as catalysts as well as surface state passivating oxides. As in the case of $BiVO_4$ (Fatwa et al., 2013), the interaction of a catalyst deposited and annealed on top of the photoelectrode metal oxide can lead to an improvement but also to a degeneration of the photoelectrode's activity.

An especially interesting class of materials are perovskite- and ferrite-type compounds, but also other multinary oxides and oxynitrides characterized by a distortion of the crystal lattice leading to ferroelectric effects are of interest (Grinberg et al., 2013). The spontaneous electric polarization can help promote the separation of photoexcited electron–hole pairs which occur at different facets of the crystallites of the electrode. It is suggested that in these materials, photovoltages can be generated which are higher than the band gap of the materials. As a function of layer thickness and engineering of domain structures, the photocurrent density of the ferroelectric absorber has been demonstrated. Not only perovskites, but also other structures, showing a strong inversion symmetry breaking, are of interest, such as $BiVO_4$.

7. SUMMARY

Hybrid PECs consisting of a low-cost PV cell of high durability and a corrosion stable photoanode arranged in a monolithic device are presently most promising systems related to technical and economic feasibility. The photoanode materials and their preparation techniques, which are presently under investigation, have to be further improved to optimize the optoelectronic properties of the absorber materials and their interfaces as well as the activity of the electrocatalysts. A crucial role plays the recombination of minority charge carriers at the absorber/window and the photoanode/electrolyte interface. For that reason, a detailed knowledge of the defect chemistry of the absorbers as a function of preparation conditions is a prerequisite to tailor the catalytic properties of the material's surface in contact with an electrolyte. On the other hand, also alternative routes using molecular approaches should be kept in mind. In all cases stability, availability, environmental sustainability and efficiency are crucial basic conditions for a useful application to convert sunlight into chemical energy.

Technologies leading to efficient conversion and storage of sunlight into chemical energy open the unique possibility of providing a renewable and sustainable energy supply based on the nearly infinite radiation energy of the sun. Since the availability of renewable energy sources is highly fluctuating depending on the presence of sunlight or wind, converted energy has to be stored.

The converted energy can be used as a power source for all kinds of applications with no constraints in time. For this purpose, it is envisioned to develop, for example, a 'membrane' which, once immersed in an aqueous electrolyte, is able to split water into hydrogen and oxygen when illuminated by sunlight. Such a device has been addressed as 'artificial leaf'. It should consist of cheap, abundant and nontoxic elements featuring a solar-to-hydrogen efficiency $>10\%$ and achieve a price of <3 €/kg H_2. Fuels, which are necessary to provide mobility, can be obtained by converting the hydrogen harvested from sunlight into hydrocarbons in a second step.

Among biomimetic approaches using inorganic systems, one possible solution is the conversion of sunlight into chemical energy via photonic excitation of a photovoltaic thin film, directly combined with corrosion-stable layers at its front and back surface of electrocatalysts allowing water electrolysis at the electrode—electrolyte interfaces. To increase the catalytic activity of these surfaces, microstructured composites and semiconducting materials belonging to the class of the ferroelectrics are of special interest. Generated hydrogen can be stored as compressed gas, as liquid H_2, as metal hydride or as methanol. Alternatively, CO_2 reduction and hydrocarbon production can be achieved on the cathode side of the device. In both cases, noble metal-free catalysts are needed to develop electrodes for a mass market with respect to oxidation/reduction of water and reduction of CO_2, respectively.

REFERENCES

Ager, J. W., Shaner, M. R., Walczak, K. A., Sharp, I. D., & Ardo, S. (2015). Experimental demonstrations of spontaneous, solar-driven photoelectrochemical water splitting. *Energy and Environmental Science, 8*, 2811—2824.

Ayers, K. E., Danilovic, N., & Anderson, E. (2014). *Challenges and limitations of materials for membrane-based water electrolysis at megawatt scale*. Abstract # 1512, 227th ECS Meeting Chicago.

Barber, J. (2012). *Solar fuels and artificial photosynthesis — Science and innovation to change our future energy options*. www.rsc.org/sustainable-energy.

Behrens, M., Studt, F., Kasatkin, I., Kühl, S., Hävecker, M., Abild-Pedersen, F., & Schlögl, R. (2012). The active site of methanol synthesis over $Cu/ZnO/Al_2O_3$ industrial catalysts. *Science, 336*, 893–897.

Bogdanoff, P., Stellmach, D., Gabriel, O., Stannowski, B., Schlatmann, R., van de Krol, R., & Fiechter, S. (2015). Artificial leaf for solar water splitting based on a triple–junction a-Si-μc-Si thin film solar cell and a PEDOT: PSS/catalyst blend. *Energy Technology, 4*, 230–241.

Brendli, J. (2011). *Storage of wind power and solar power within the existing gas grids as a basic concept in the change of energy supply towards renewable energy – Demand and comprehension of accumulating energy* (Bachelor thesis, Hochschule München/Thüga AG).

Brillet, J., Yum, J. H., Cornuz, M., Hisatomi, T., Solarska, R., Augustynski, J., … Sivula, K. (2012). Highly efficient water splitting by a dual-absorber tandem cell. *Nature Photonics*. http://dx.doi.org/10.1038/NPHOTON.2012.265.

Bundesministerium für Wirtschaft und Energie. (2015). *Primärenergieverbrauch in Deutschland*. www.bmwi.de/DE/Themen/Energie/konventionelle-energietraeger,did=540366.html.

Ciamician, G. (1912). The photochemistry of the future. *Science, 36*, 385–394.

Dong, W., Guo, Y., Guo, B., Li, H., Liu, H., & Joel, T. W. (2013). Enhanced photo voltaic effect in $BiVO_4$ semiconductor by incorporating with an ultrathin $BiFeO_3$ ferroelectric layer. *ACS Applied Materials and Interfaces, 5*, 6925–6929.

Fatwa, F. A., Han, L., Smets, A. H. M., Zeman, M., Dam, B., & van de Krol, R. (2013). Efficient solar water splitting by enhanced charge separation in a bismuth vanadate-silicon tandem photoelectrode. *Nature Communications*. http://dx.doi.org/10.1038/ncomms3195.

Fiechter, S., Bogdanoff, P., Bak, T., & Nowotny, J. (2012). Basic concepts of photo electrochemical solar energy conversion systems. *Advances in Applied Ceramics, 111*, 39–43.

Fujishima, A., & Honda, K. (1972). Electrochemical photolysis of water at a semiconductor electrode. *Nature, 238*, 37–38.

Grinberg, I., West, D. V., Torres, M., Gou, G., Stein, D. M., Wu, L., & Rappe, A. M. (2013). Perovskite oxides for visible-light-absorbing ferroelectric and photovoltaic materials. *Nature, 503*, 509–512.

Hori, Y. (2012). CO_2-reduction, catalyzed by metal electrodes. In *Handbook of fuel cells*. New York: Wiley. http://dx.doi.org/10.1002/9780470974001.f207055.

Jacobsson, T. J., Fjallstrom, V., Sahlberg, M., Edoff, M., & Edvinsson, T. A. (2013). Monolithic device for solar water splitting based on series interconnected thin film absorbers reaching over 10% solar-to-hydrogen efficiency. *Energy and Environmental Science, 6*, 3676–3683.

Kaiser, B., Jaegermann, W., Fiechter, S., & Lewerenz, H. J. (2011). Direct photo electrochemical conversion of sun light into hydrogen for chemical energy storage. *Bunsen-Magazin, 13*, 104–111.

Khaselev, O., & Turner, J. A. (1998). A monolithic photovoltaic-photoelectrochemical device for hydrogen production via water splitting. *Science, 280*, 425–427.

Kim, J. Y., Magesh, G., Youn, D. H., Jang, J.-W., Kubota, J., Domen, K., & Lee, J. S. (2013). Single-crystalline, wormlike hematite photoanodes for efficient solar water splitting. *Nature*. http://dx.doi.org/10.1038/srep02681.

Kondratenko, E. V., Mul, G., Baltrusaitis, J., Larrazábal, G. O., & Pérez-Ramírez, J. (2013). Status and perspectives of CO_2 conversion into fuels and chemicals by catalytic, photocatalytic and electrocatalytic processes. *Energy and Environmental Science, 6*, 3112–3135.

Lawrence Livermore National Laboratory, Energy Flow, Archive for Charts: https://flowcharts.llnl.gov/archive.html, https://flowcharts.llnl.gov/archive.html, https://flowcharts.llnl.gov/content/energy/energy_archive/energy_flow_2013/2013USEnergy.png.

Lewinski, K. A., & Luope, S. M. (2014). *High power water electrolysis as a new paradigm for operation of PEM electrolyzer*. Abstract # 1948, 227th ECS Meeting Chicago.

Li, H., Lei, Y., Huang, Y., Fang, Y., Xu, Y., Zhu, L., & Li, X. (2011). Photocatalytic reduction of carbon dioxide to methanol by Cu_2O/SiC. Nanocrystallite under visible light irradiation. *Journal of Natural Gas Chemistry, 20*, 145–150.

Li, R., Zhang, F., Wang, D., Yang, J., Li, M., Zhu, J., … Li, C. (2013). Spatial separation of photogenerated electrons and holes among {010} and {110} crystal facets of $BiVO_4$. *Nature Communications*. http://dx.doi.org/10.1038/ncomms2401.

Licht, S., Wang, B., Mukerji, S., Soga, T., Umeno, M., & Tributsch, H. (2000). Efficient solar water splitting, exemplified by RuO_2-catalyzed AlGaAs/Si photo electrolysis. *Journal of Physical Chemistry B, 104*, 8920—8924.

Liu, Y., Huang, B., Dai, Y., Zhang, X., Qin, X., Jiang, M., & Whangbo, M. W. (2009). Selective ethanol formation from photocatalytic reduction of carbon dioxide in water with $BiVO_4$ photocatalyst. *Catalysis Communications, 11*, 210—213.

Luo, J., Im, J. H., Mayer, M. T., Schreier, M., Nazeeruddin, M. K., Park, N. G., … Grätzel, M. (2014). Water photolysis at 12.3% efficiency via perovskite photovoltaics and Earth-abundant catalysts. *Science, 345*. http://dx.doi.org/10.1126/science.1258307.

Marcel Schreier, M., Curvat, L., Giordano, F., Steier, L., Abate, A., Zakeeruddin, S. M., … Grätzel, M. (2015). Efficient photosynthesis of carbon monoxide from CO_2 using perovskite photovoltaics. *Nature Communications*. http://dx.doi.org/10.1038/ncomms8326.

May, M. M., Lewerenz, H.-J., Lackner, D., Dimroth, F., & Hannappel, T. (2015). Direct solar-to-hydrogen conversion by in situ interface transformation of a tandem structure. *Nature Communications*. http://dx.doi.org/10.1038/ncomms9286.

McCrory, C. C. L., Jung, S., Ferrer, I. M., Chatman, S. M., Peters, J. C., & Jaramillo, T. F. (2015). Benchmarking hydrogen evolving reaction and oxygen evolving reaction electrocatalysts for solar water splitting devices. *Journal of the American Chemical Society, 137*, 4347—4357.

McCrory, C. C. L., Jung, S., Peters, J. C., & Jaramillo, T. F. (2013). Benchmarking heterogeneous electrocatalysts for the oxygen evolution reaction. *Journal of the American Chemical Society, 2013*(135), 16977—16987.

Morales-Guio, C. G., Tilley, S. D., Vrubel, H., Grätzel, M., & Hu, X. (2014). Hydrogen evolution from a copper(I) oxide photocathode coated with an amorphous molybdenum sulphide catalyst. *Nature Communications*. http://dx.doi.org/10.1038/ncomms4059.

Nakamura, A., Ota, Y., Koike, K., Hidaka, Y., Nishioka, K., Sugiyama, M., & Fujii, K. (2015). A 24.4% solar to hydrogen energy conversion efficiency by combining concentrator photovoltaic modules and electrochemical cells. *Applied Physics Express, 8*, 107101-1—107101-4.

Paracchino, A., Laporte, V., Sivula, K., Grätzel, M., & Thimsen, E. (2011). Highly active oxide photocathode for photoelectrochemical water reduction. *Nature Materials, 10*, 456—461.

Peharz, G., Dimroth, F., & Wittstadt, U. (2007). Solar hydrogen production by water splitting with a conversion efficiency of 18%. *International Journal of Hydrogen Energy, 32*, 3248—3252.

Pihosh, Y., Turkevych, I., Mawatari, K., Uemura, J., Kazoe, Y., Kosar, S., … Kitamori, T. (2015). Photocatalytic generation of hydrogen by core-shell $WO_3/BiVO_4$ nanorods with ultimate water splitting efficiency. *Nature*. http://dx.doi.org/10.1038/srep11141.

Rau, S., Vierrath, S., Ohlmann, J., Fallisch, A., Lackner, D., Dimroth, F., & Smolinka, T. (2014). Highly efficient solar hydrogen generation — an integrated concept joining III-V solar cells with PEM electrolysis cells. *Energy Technology, 2*, 43—53 (and references therein).

Reece, S. Y., Hamel, J. A., Sung, K., Jarvi, T. D., Esswein, A. J., Pijpers, J. J. H., & Nocera, D. G. (2011). Wireless solar water splitting using silicon-based semiconductors. *Science, 334*, 645—648.

Shaner, M. R., Fountaine, K. T., Ardo, S., Coridan, R. H., Atwater, H. A., & Lewis, N. S. (2013). Photoelectrochemistry of core—shell tandem junction $n-p^+$-Si/n-WO_3 microwire array photoelectrodes. *Energy and Environmental Science*. http://dx.doi.org/10.1039/c3ee43048k (and references therein).

Stellmach, D. (2014). *Hybrid-PV-Elektroden zur Wasserspaltung funktionalisiert mit MoS_x-Nanopartikeln als Wasserstoff-entwickelndem Elektrokatalysator* (Ph.D. thesis, TU-Berlin).

Stellmach, D., Bogdanoff, P., Gabriel, O., Stannowski, B., Schlatmann, R., van de Krol, & Fiechter, S. (2013). Nanostructured MoS_2 particles as a novel hydrogen evolving catalyst integrated in a PV-hybrid electrolyzer. In A. Méndez-Vilas (Ed.), *Materials and processes for energy: Communicating current research and technological developments* (pp. 880–886). FORMATEX.

Styring, S. (2015). Artificial photosynthesis for solar fuels. *Faraday Discussions, 2012*(155), 357–376. www.kemi.uu.se/research/molecular-biomimetic/photosynthesis/styring-group/ongoing-research/artificial-photosynthesis.

Urbain, F., Smirnov, V., Becker, J.-P., Lambertz, A., Yang, F., Ziegler, J., … Finger, F. (2015). Multijunction Si photocathodes with tunable photovoltages from 2.0 V to 2.8 V for light induced water splitting. *Energy and Environmental Science*. http://dx.doi.org/10.1039/c5ee02393a.

Wirth, H. (2015). *Aktuelle Fakten zur Photovoltaik in Deutschland*. www.ise.fraunhofer.de/de/veroeffentlichungen/studien-und-positionspapiere/aktuelle-fakten-zur-photovoltaik-in-deutschland.

Yang, X., Du, C., Du, C., Liu, R., Xie, J., & Wang, D. (2013). Balancing photovoltage generation and charge-transfer enhancement for catalyst-decorated photoelectron chemical water splitting: a case study of the hematite/MnO_x combination. *Journal of Catalysis, 304*, 86–91.

Yui, T., Tamaki, Y., Sekizawa, K., & Ishitani, O. (2011). Photocatalytic reduction of CO_2: from molecules to semiconductors. In A. Bignozzi (Ed.), *Photocatalysis (Topics in current chemistry 303)* (pp. 151–184). Heidelberg: Springer.

Zhong, D. K., & Gamelin, D. R. (2010). Photoelectrochemical water oxidation by cobalt catalyst ("Co-Pi")/r-Fe2O3 composite photoanodes: oxygen evolution and resolution of a kinetic bottleneck. *Journal of the American Chemical Society, 132*, 4202–4207.

CHAPTER FIVE

Artificial Photosynthesis: Theoretical Background

D. Abramavicius*, L. Valkunas*,§,1
*Vilnius University, Vilnius, Lithuania
§Center for Physical Sciences and Technology, Vilnius, Lithuania
[1]Corresponding author: E-mail: leonas.valkunas@ff.vu.lt

Contents

Abstract

Molecular aggregates and their specific structural arrangements are used in constructing various artificial light conversion devices mimicking the natural photosynthesis. Review of several theoretical approaches applicable for the description of the excitation transfer and charge separation in these systems is presented. From the theoretical point of view two competing processes, namely, excitation resonance transfer between constituent molecules and their interaction with intra- and intermolecular vibrations (and phonons of the environment), are essential factors determining the fate of electronic excitations. Possible scenarios determining the excitation dynamical evolution are described in the chapter. For demonstration purposes the excitation dynamics in a molecular heterodimer is briefly discussed. The heterodimer under consideration is

Advances in Botanical Research, Volume 79
ISSN 0065-2296
http://dx.doi.org/10.1016/bs.abr.2016.04.001

assumed to be composed of two molecules modelled as two-level systems coupled by the resonance interaction. The system interaction with phonon bath is also taken into account. The system is studied as a function of difference in excitation and reorganization energies, the asymmetry in transition dipole moments and excited-state lifetimes. This model demonstrates the limitations of various theoretical approaches.

1. INTRODUCTION

Large molecules absorbing solar irradiation in the visible region and their aggregates are core constituents of light accumulation/harvesting and conversion systems of natural and artificial origin (van Amerongen, Valkunas, & van Grondelle, 2000; Pope & Swenberg, 1999). In natural photosynthetic systems they often arrange into the excitation transfer networks thus enabling solar energy funnelling to specific energy conversion units (van Amerongen et al., 2000). Acceptance and donation ability of electrons in such systems leads to interconversion of excitons and mobile charge carriers responsible for energy conversion in photosynthesis as well as in various organic optoelectronic devices, like solar cells. Interest in excitation dynamics notably increased during the past years due to detailed knowledge of the naturally occurring molecular structures, for example, of photosynthetic aggregates (Blankenship, 2014; Ruban, 2012), and due to the capability of synthesizing molecular assemblies of predefined architecture (O'Sullivan et al., 2011) (for the review see chapters of this book). In the latter case the concept of the structural organization is based on the approach of mimicking the natural photosynthesis.

To describe the excitation evolution after photon absorption in molecular aggregates the density matrix formalism developed for the description of perturbation dynamics in open quantum systems is usually exploited (Breuer & Petruccione, 2002; van Kampen, 2007; May & Kühn, 2011; Valkunas, Abramavicius, & Mancal, 2013; Weiss, 2008). In general, the full-density matrix of a quantum system interacting with the environment is described by the Liouville equation:

$$\frac{d\widehat{W}}{dt} = -\frac{i}{\hbar}\mathscr{L}\widehat{W}, \tag{1}$$

where $\mathscr{L}$ is the Liouville superoperator, given by a commutator with the total Hamiltonian:

$$\mathscr{L}\widehat{W} = \left[\widehat{H}, \widehat{W}\right]. \tag{2}$$

The density operator describes the complete dynamics of the closed system. However, such description is possible only for a limited set of quantum problems, and approximate methods are usually involved. For this purpose, the closed system is split into two or several *system* parts which are treated completely, while the remaining degrees of freedom are taken as perturbations. Specific chosen part of the system constitutes the basis set of the quantum problem formulation. At this point the description is usually based on two limiting cases. In the Redfield approach the system part includes only electronic degrees of freedom, while vibrational degrees of freedom (inter- and intramolecular vibrations, phonons etc.) are taken as perturbations performing the thermal Brownian fluctuations. In the Förster description the distinction is more subtle: the system consists of molecular electronic transitions of different molecules together with the nuclear vibrations causing the transition energy modulation (here the nuclear coordinates of each molecule are different and independent); the resonance coupling elements between the molecules are treated pertubatively. For the description of the electron dynamics the third, Marcus, theory assumes the strong coupling of an electron under consideration with one or few nuclear coordinates. In this case the same nuclear coordinate is assumed to be coupled with the two, initial and final, electronic states. The detailed consideration of the energy or charge transfer can be obtained by following the coupling between electronic and vibrations subsystems and the influence of the coherent vibrations on the exciton/electron dynamics can be revealed.

2. GENERIC HAMILTONIAN OF THE COUPLED SYSTEM

Frenkel exciton model is the primary construction when describing an excitation in the system of coupled molecules. Electronically interacting pigment molecules can be well described by means of the Heitler–London approximation (van Amerongen et al., 2000; Davydov, 1962; Valkunas et al., 2013). In this model the global ground state is taken to have zero energy. The electronic excited state of the aggregate is defined as the electronic excitation of a single molecule when all other molecules are in their ground states. The electronic system is thus defined by the following Hamiltonian:

$$\widehat{H}_{\text{exc}} = \sum_{m}(\varepsilon_m + \delta\varepsilon_m)\widehat{B}_m^{\dagger}\widehat{B}_m + \sum_{m\neq n} J_{mn}\widehat{B}_m^{\dagger}\widehat{B}_n. \qquad (3)$$

Here electronic excitation energies are ε_m. $\delta\varepsilon_m$ is the stochastic excitation energy offset later used to incorporate the static disorder of site energies. J_{mn} is the intermolecular resonant coupling matrix. Operator $\widehat{B}_m^\dagger$ defines the electronic excitation of the molecule m.

The molecules are usually accepted as two-level systems with respect to electronic states, that is, they are either excited or unexcited. In this case a molecule cannot be excited twice. Consequently, the operators satisfy specific 'Pauli' commutation relations, $[\widehat{B}_m, \widehat{B}_n^\dagger] = \delta_{mn}(1 - 2\widehat{B}_m^\dagger \widehat{B}_m)$.

With the restriction of a single excitation this Hamiltonian describes a square matrix, which after diagonalization yields excitonic delocalized eigenstates. However, so far the properties of molecules are not completely specified since nuclear degrees of freedom have been neglected. To include molecular vibrations the multidimensional displaced harmonic oscillator model can be used for description of excitation dynamics in each pigment molecule. In this model, the potential energy surfaces of the excited and ground states are approximated as harmonic potentials along general nuclear coordinates (usually normal modes) q_m, which are defined in the reference frame of the mth molecule (Basinskaite, Butkus, Abramavicius, & Valkunas, 2014; Butkus, Valkunas, & Abramavicius, 2014). In one-dimensional case the total energy of the ground state of a single-harmonic mode of mth molecule is then defined as follows:

$$\widehat{h}_m(\widehat{q}_m) = \frac{1}{2M}\left(M^2\omega_m^2\widehat{q}_m^2 + \widehat{p}_m^2\right), \tag{4}$$

where $\widehat{p}_m$ is a canonical momentum operator and ω_m is the oscillator (vibrational) frequency. M defines the oscillating mass. The nuclear modes can be quantized using bosonic creation/annihilation operators $\widehat{a}_m^\dagger$ and $\widehat{a}_m$, respectively, yielding

$$\widehat{h}_m(\widehat{q}_m) = \hbar\omega_m\left(\widehat{a}_m^\dagger\widehat{a}_m + \frac{1}{2}\right). \tag{5}$$

Now the electronic basis of coupled molecules consists of the common electronic ground state $|0\rangle$ and molecular excited states $|m\rangle$, which can be related by using the electronic excitation creation operator $\widehat{B}_m^\dagger$: $|m\rangle = \widehat{B}_m^\dagger|0\rangle$.The nuclear potential surface gets shifted in the electronic excited state. Notice here the difference between the electronic ground state $|0\rangle$ which can be vibrationally excited and the complete vacuum state $|\varnothing\rangle$. In the Hamiltonian of molecular aggregates, the remaining terms represent resonance interaction between molecular transitions:

$$\begin{aligned}\widehat{H}_{\mathrm{MA}} = &\sum_m \widehat{h}_m(\widehat{q}_m) + \sum_m \left[\varepsilon_m + \delta\varepsilon_m + \widehat{h}_m(\widehat{q}_m + d_m)\right]\widehat{B}_m^\dagger\widehat{B}_m \\ &+ \sum_{m\neq n} J_{mn}\widehat{B}_m^\dagger\widehat{B}_n.\end{aligned} \tag{6}$$

analogously as in Eq. (6), and the excited state is not displaced ($|m\rangle = \widehat{B}_m^{\dagger}|0\rangle$):

$$\widehat{H}'_{\mathrm{MA}} = \sum_m \hbar\omega_m\left(\widehat{a}_m^{\dagger}\widehat{a}_m + \frac{1}{2}\right) + \sum_{m \neq n} J_{mn}\widehat{B}_m^{\dagger}\widehat{B}_n + \sum_m \left[\varepsilon_m + \delta\varepsilon_m + \lambda_m - \hbar\omega_m\sqrt{s_m}\left(\widehat{a}_m + \widehat{a}_m^{\dagger}\right)\right]\widehat{B}_m^{\dagger}\widehat{B}_m. \tag{14}$$

The model is completely equivalent to the shifted basis model from the physical point of view, despite that the excited-state matrices are different. In the present model the diagonal blocks $\langle e_m, (j_1 \ldots j_N)|\widehat{H}_{\mathrm{MA}}|e_m, (i_1 \ldots i_N)\rangle$, where now

$$|e_m, (i_1 \ldots i_N)\rangle = \widehat{B}_m^{\dagger}\prod_n \frac{\left(\widehat{a}_n^{\dagger}\right)^{i_n}}{\sqrt{i_n!}}|\varnothing\rangle, \tag{15}$$

have off-diagonal elements, which are due to the $\hbar\omega_0\sqrt{s_m}(\widehat{a}_m + \widehat{a}_m^{\dagger})\widehat{B}_m^{\dagger}\widehat{B}_m$ term in the Hamiltonian; the off-diagonal blocks are, however, diagonal because the intermolecular couplings do not have shift operators.

Differences also appear in expressions of the transition dipole moment. In the ground-state basis the dipole moment for the aggregate is given by

$$\widehat{\overrightarrow{\mu}} = \sum_m \overrightarrow{d}_m\widehat{B}_m^{\dagger} + \mathrm{h.c.} \tag{16}$$

The dipole moment matrix element in this basis is

$$\overrightarrow{\mu}_{(e_m, i_1 \ldots i_N),(g, j_1 \ldots j_N)} = \overrightarrow{d_m}\prod_n \delta_{i_n j_n} \tag{17}$$

The transformation to the eigenstate basis is equivalent as in shifted basis Hamiltonian. Both shifted models behave differently when carrying out truncated numerical calculations; however, they lead to the same results if the number of vibrational levels is taken to infinity.

2.3 Exciton–Phonon Interaction

Excitations in the molecular aggregates are usually considered as Frenkel excitons delocalized over several molecules. While the molecules may be identical in their chemical composition, the conformational disorder breaks the symmetries. Additionally, molecular vibrations perturb the energies of electronic transitions. Thus we apply the model including the static and time-independent site-energy variation induced by overdamped or moderately damped phonons (Valkunas et al., 2013):

$$\hat{H}_{\mathrm{B}} + \hat{H}_{\mathrm{SB}} = \sum \hbar_\alpha \omega_\alpha \hat{b}_\alpha^\dagger \hat{b}_\alpha - \sum_n \hat{B}_n^\dagger B_n \sum_q g_{n\alpha} \hbar \omega_\alpha \left(\hat{b}_\alpha^\dagger + \hat{b}_\alpha \right). \quad (18)$$

Here $\hat{b}_\alpha^\dagger$ and $\hat{b}_\alpha$ are the creation and annihilation operators for the vibrational (phonon field) mode, α is the mode index, ω_α is its frequency, and $g_{n\alpha}$ is the linear coupling strength between the site n and the phonon mode α.

These modes will be treated approximately since their number is infinite. Hence it is impossible to include them explicitly as is the case of the molecular normal modes, which are described earlier. Thus, Eq. (18) describes perturbative fluctuations to the Hamiltonian. However, it is usually convenient to represent the dissipative dynamics in the eigenbasis of system degrees of freedom. In that case the excitonic Hamiltonians need to be diagonalized. Consequently, let us assume that the latter step has already been performed and the system Hamiltonian is diagonal. The system—bath coupling matrix can then be transformed into the eigenbasis. This leads to emergence of fluctuating couplings (off-diagonal fluctuations) which are dynamically induced by environment fluctuations. Additionally, the fluctuations create fluctuations of energies (diagonal fluctuations). It is shown by Redfield theory (see Section 3.1) that the diagonal and off-diagonal fluctuations cause different physical phenomena: pure dephasing and energy relaxation, respectively.

3. MASTER EQUATIONS FOR THE DENSITY MATRIX

3.1 Redfield Relaxation Theory

All Hamiltonians presented earlier can be split into three terms: the isolated system part, the environment (nuclear) part and the cross term or the coupling between the system and the environment. The total supersystem, which consists of both ingredients (the observable system and the environment), is still closed. Within the frame of the Redfield theory it is a common practice to expand the right hand side (RHS) of the Liouville equation, Eq. (1), in powers of the coupling term up to second order and take the trace operation over the equilibrium environment (canonical ensemble average). The resulting thermally averaged density matrix is then denoted as the reduced density matrix depending only on the degrees of freedom of the observable system $\widehat{\rho} = Tr_{\mathrm{E}}(\widehat{W})$ (E indicates the environment), however, the equation of motion indirectly involves the degrees of freedom of the environment through the relaxation superoperators, which will be denoted by $\mathscr{R}$.

The relaxation superoperators involve the two-point quantum correlation functions, $C(t)$, of the environment coordinates. In a commonly used approach the environment is taken as a collection of harmonic oscillators, indexed by α, and, thus, the correlation functions of the form $C(t) \sim \sum_{\alpha} k_{\alpha} \langle q_{\alpha}(t) q_{\alpha}(0) \rangle$ have analytic expressions. Here q_{α} is the coordinate operator of the environment oscillator in the Heisenberg representation. At constant temperature using the canonical ensemble formulation they imply the Brownian fluctuations (Valkunas et al., 2013). The correlation function is usually a decaying function, characterized by a timescale τ_{c} — the correlation decay time. It describes the environment relaxation/dephasing time (notice that it is different from the system relaxation/dephasing time).

The environment coordinates can be explicitly associated with the intermolecular modes (or phonons) and the weakly coupled intramolecular modes (or the molecular vibrations). Taking this into account we may observe very complicated set of environment oscillators. Distribution of their frequencies may consist of many peaks. A function denoted by a fluctuation spectral density, $C''(\omega) = -2\Im \int_0^{\infty} \sin\omega t \cdot C(t) dt$ is a characteristic function modelling the environment. It is usually a multipeak function with a smooth background (Butkus, Valkunas, & Abramavicius, 2012, May & Kühn, 2011). The peaks reflect the intramolecular vibrations unaccounted in Sections 2.1 and 2.2, while the background characterizes phonons. This function or its another form, $J(\omega) = C''(\omega)/\omega^2$, is also an experimentally measurable function defined from, for example, fluorescence line narrowing spectra (Freiberg, Lin, Timpmann, & Blankenship, 1997; Jankowiak, Reppert, Zazubovich, Pieper, & Reinot, 2011; Kell, Feng, Reppert, & Jankowiak, 2013; Timpmann, Rätsep, Hunter, & Freiberg, 2004) or the coherent two-dimensional electronic spectroscopy (Rancova, Jankowiak, & Abramavicius, 2015).

The coupling between the system and the environment causes the fluctuations of the observable system parameters. In the matrix representation the diagonal and off-diagonal fluctuations of the Hamiltonian terms can be distinguished and separated. If the system is represented by its eigenstates, such separation is very convenient since their influence becomes straightforward: the diagonal fluctuations cause pure dephasing of system coherences, while the off-diagonal fluctuations cause the population transport/relaxation. Notice that any other representation can be transformed into a chosen basis set where the Hamiltonian of interest is diagonal. The concept of correlation functions and spectral densities directly extends to classical dimensions, where the coordinates q_α perform the Brownian fluctuations. The stochastic dynamical formulation can thus be employed (Abramavicius & Abramavicius, 2014; Chorosajev, Gelzinis, Valkunas, & Abramavicius, 2014; van Kampen, 2007; Valkunas et al., 2013).

In the case of weak system-environment coupling the environment degrees of freedom are expanded up to the second order in the Liouville equation giving the Redfield equation (or the Redfield theory), which describes the exciton dynamics/relaxation in terms of the master equations for the reduced density matrix (Redfield, 1957). In that case the coherent dynamics of the system is assumed to decay slower than the environment correlation functions, providing also the applicability of the Markovian approximation, which allows the Redfield equation to be presented in a very simple form:

$$\frac{d\rho_{ij}}{dt} = -i\omega_{ij}\rho_{ij} - \mathcal{R}_{ij,kl}\rho_{kl}. \tag{19}$$

Here $\hbar\omega_{ij} = \varepsilon_i - \varepsilon_j$ is the system energy gap, which describes the coherent system frequencies, while $\mathcal{R}$ is the relaxation superoperator tetradic matrix. Mathematically this equation is a linear complex-valued master equation. The diagonal elements ρ_{ii} describe the occupation probabilities for states i and the off-diagonal elements ρ_{ij} describe the coherences.

When the energy gaps $\hbar\omega_{ij}$ are larger than the relaxation rates, a secular approximation is valid as well. In this case the relaxation matrix is 'diagonal' in the sense that the nonzero elements are only $\mathcal{R}_{ii,jj}$ and $\mathcal{R}_{ij,ij}$; the former describe the population relaxation, while the latter are the coherence decay rates. Consequently, the relaxation superoperator is reduced to two blocks and neglects the mixing between the coherences and the populations. In this approximation the population dynamics is described by the master equation of the form

$$\frac{d\rho_{ii}}{dt} = \sum_{j}^{j\neq i} k_{ij}\rho_{jj} - k_{ii}\rho_{ii}, \tag{20}$$

where $k_{ij} = -\mathcal{R}_{ii,jj}$ for $i \neq j$ and $k_{ii} = \mathcal{R}_{ii,ii}$. The rate constants k_{ij} characterize the population flow from state j into state i. According to the model of the harmonic bath it is given by (Valkunas et al., 2013)

$$k_{ij} = 2C''_{ij}(\omega_{ij})\frac{\exp\left(\frac{\hbar\omega_{ij}}{k_B T}\right)}{\exp\left(\frac{\hbar\omega_{ij}}{k_B T}\right) - \exp\left(-\frac{\hbar\omega_{ij}}{k_B T}\right)}, \tag{21}$$

Exponential factor ensures that the down-hill transfer is activationless while the up-hill transfer requires thermal activation in order to supply energy for electronic states from the environment. The spectral density $C''_{ij}(\omega)$ characterizes the off-diagonal Hamiltonian fluctuations between states i and j, $J_{ij}(t) \sim \sum_{\alpha} c_{ij\alpha} q_{\alpha}(t)$ (notice that in the eigenstate basis $J_{ij}(t)$ on average is zero):

$$C''_{ij}(\omega) = -2\Im \int dt \sin(\omega t)\left\langle J_{ij}(t) J_{ji}(0)\right\rangle_B. \tag{22}$$

Consequently, from Eqs (21) and (22) it follows that the fluctuations of the *environment* are promoting the $j \to i$ transition of the system through rescaled spectral densities of the coordinate fluctuations. Since the spectral density enters at the specific frequency ω_{ij}, it is necessary that there must be a vibrational mode containing frequencies matching the energy gap $\hbar\omega_{ij}$ between the system states. Hence, the Redfield theory describes the population transfer promoted by vibrational resonances, where the perfect match of electronic gaps and vibrational frequencies is necessary. This matching ensures the energy conservation. The energy relaxation or the energy loss is possible whenever there is the energy accepting compartment.

According to the Redfield theory, the nuclear motion is always at equilibrium, which means that the nuclei affect the electronic motion, but not vice versa. This applies for the nuclear coordinates q_{α}, which experience the thermal motion, thus they are neither modulated nor magnified by changes in electronic configurations. However, the fluctuations of Hamiltonian elements are scaled by the coupling strengths $g_{n\alpha}$. Hence the change in electronic configuration J_{ij} or J_{kl} will show different degree of nuclear dynamics. One more important observation is that the electronic coherence ρ_{ij} is not coupled to the populations and, thus, the electronic coherence does not affect the population relaxation.

The equations for coherences are given by

$$\frac{d\rho_{ij}}{dt} = -i\omega_{ij}\rho_{ij} - \gamma_{ij}\rho_{ij}, \tag{23}$$

where $\gamma_{ij} = \mathscr{R}_{ij,ij}$. It turns out that the dephasing rate γ_{ij} is dependent on the population rates k_{ij} in a simple way: $\gamma_{ij} = (k_{ii} + k_{jj})/2 + \overline{\gamma}_{ij}$ (Valkunas et al., 2013). The first term is related to the lifetime of the states i and j; the second term — the pure dephasing — is due to diagonal fluctuations of states i and j, that is, $J_{ii}(t)$ and $J_{jj}(t)$. As shown before, the population rates demonstrate the resonance behaviour between the system and the environmental frequencies — there is no transfer in the absence of resonance, however, the pure dephasing term is completely off-resonant, it is given by the total integral of the correlation function and is not related to the environmental frequencies.

However, without the secular approximation the electronic coherences become involved in the population relaxation and the *quantum transport* may take place (Eq. (19)). In this case the populations obtain the oscillatory contributions through the population—coherence mixing terms:

$$\frac{d\rho_{ii}}{dt} = \ldots - \mathcal{R}_{ii,kl}\rho_{kl} - \mathcal{R}_{ii,lk}\rho_{lk} \tag{24}$$

when $\rho_{kl} \approx \exp(-i\omega_{kl}t)$. Additionally, the coherences become driven by the populations:

$$\frac{d\rho_{ij}}{dt} = -i\omega_{ij}\rho_{ij} - \ldots\mathcal{R}_{ij,kk}\rho_{kk}. \tag{25}$$

The experimental manifestation of the quantum transport was proposed during 2010s (Abramavicius & Mukamel, 2010; Fuller et al., 2014; Panitchayangkoon et al., 2011). It should be noted that the conditions of the secular approximation can be violated in the photosynthetic pigment protein aggregates as well as in molecular systems mimicking the photosynthetic construction. So while the electronic coherence seems to play a minor role in energy transport in this formulation, its role should not be underestimated for the very existence of the transfer rates $\mathcal{R}_{ij,kl}$.

Thus, it is noteworthy that the Redfield theory describes quantum relaxation. The energy and electron transport may not be directly inferred from the classical master equation, which describes only the population relaxation. However, it must be taken into account that these populations correspond to delocalized quantum states, which in the real space are superpositions of molecular excited states. So while their description appears to obey the classical equation in the eigenstate basis, the essential dynamics is wavelike in the real space and the energy flow through the system may be inferred by following the exciton wavefunction amplitude squares in the real space, $|\psi_{\text{exc,site}}|^2$.

3.2 Förster Energy Transfer

This model of relaxation originally describes the incoherent excitation population transfer or hopping and is usually named as the FRET regime (Förster, 1948). However, this theory has been extended to include coherences as well (Balevičius, Gelzinis, Abramavicius, Mančal, & Valkunas, 2012; Balevičius, Gelzinis, Abramavicius, & Valkunas, 2013; Mančal, Balevičius, & Valkunas, 2011), which finally results in the formation of localized polaronic excitations (Gelzinis, Abramavicius, & Valkunas, 2011). In the Förster description the system is assumed to have electronic quantum states, which are directly associated with separate molecules in a real space. The molecular transition energies experience diagonal fluctuations induced by nuclear vibrations. Additionally, the molecules are weakly coupled to each other. Again all degrees of freedom can be similarly partitioned into two parts; however, in this case the static intermolecular coupling is taken as a perturbation. Thus, it is possible to obtain the population transfer rate expression by including the diagonal nuclear fluctuations in terms of the cumulant expansion technique, giving the exact results for the harmonic bath, while the off-diagonal couplings remain to be at the second order. The Förster theory gives the excitation transfer rate from the initially excited donor molecule to the final acceptor molecule. The resulting rate expression is given by the convolution of the absorption $\mathscr{A}_a(\omega)$ and emission $\mathscr{F}_b(\omega)$ lineshape functions (unit area) of the final, a, and initial, b, states respectively (May & Kühn, 2011; Valkunas et al., 2013):

$$k_{ba} = \frac{2\pi}{\hbar^2}\left|J_{ab}\right|^2 \int \frac{d\omega}{2\pi} \mathscr{A}_a(\omega)\mathscr{F}_b(\omega). \tag{26}$$

It should be noted that this expression obeys the detailed balance relation as well since the pigments always absorb at higher energies than they emit due to the down-hill energy relaxation resulting in the Stokes shift.

The relation of the excitation coupling to the nuclear vibrations is obvious keeping in mind that the spectral lineshapes originate from the coupling to the nuclear vibrations/fluctuations. Consequently, the coupling with nuclear vibrations (the diagonal fluctuations) creates the conditions for the population transfer by matching the energies of the acceptor and the donor molecules. Therefore, the population transfer requires complete resonant conditions, so that the vibrational frequencies must exactly match the electronic energy gaps (alternatively there will be no overlap between the donor and acceptor spectra).

The Förster energy transfer has been generalized for the case of the two domains of chromophores instead of just two chromophores (Jang, Newton, & Silbey, 2004). This generalized Förster theory or alternatively the multichromophoric Förster theory (MC-Förster) includes the effect that the donor and acceptor domains can be delocalized excitonic states. The resulting expression is similar to the original Eq. (26); however, it now contains summation over all donor-related and acceptor-related chromophores:

$$k_{DA} = \frac{2\pi}{\hbar^2} \sum_{ik \in D} \sum_{jl \in A} J_{ij} J_{kl} \int \frac{d\omega}{2\pi} \mathscr{A}_{jl}^{(A)}(\omega) \mathscr{F}_{ik}^{(D)}(\omega), \tag{27}$$

where now $\mathscr{A}_{jl}^{(A)}(\omega)/\mathscr{F}_{ik}^{(D)}(\omega)$ are again the absorption/fluorescence spectra of the accepting/donating domains, weighted by the excitonic wavefunctions of chromophore contributions. The overall picture becomes more complicated since now the spectra are shaped by the excitonic interactions within each domain. Thus electronic intradomain coherences become involved in making spectral resonances and the vibrational bands now should match these excitonic energy gaps for the exciton interdomain transfer to be possible. It is important to note that similar to the Redfield theory, the down-hill energy transfer does not require thermal activation.

Extension of the theory to include also intermolecular quantum coherences showed that the coherences decay in a more complicated pattern than it follows from the Redfield theory (Balevičius et al., 2012, 2013; Mančal et al., 2011). An important result was obtained demonstrating that the intermolecular coupling effectively decays with time, thus showing that the *relaxed* intermolecular coupling is smaller than the original one. This describes the polaronic effect where intermolecular couplings are screened by polarization of the environment (phonons). The nuclear degrees of freedom overrule in this case the excitation trapping. The polaron formation is expected as has been described in a number of studies (Chorosajev et al., 2014; Gelzinis et al., 2011; Song & Williams, 1993).

3.3 Marcus Theory for Electron Hopping (Separation)

Relaxation theories described earlier characterize quantum transition processes which involve electronic and nuclear degrees of freedom. The electronic eigenstate energies are matched by absorption or emission of off-diagonal nuclear quanta in the Redfield theory or by energy modulation in coupled molecules within the frame of the Förster theory. Electron transfer in molecular aggregates involves large dipolar moments of the charge transfer (CT) electron–hole configurations or CT states and therefore the large reorganization energies when nuclei follow the electronic polarization. In this case it is possible to introduce the specific slow *reaction coordinate*, where the nuclear dynamics takes place as the electron configuration changes.

The Marcus theory (Barzykin, Frantsuzov, Seki, & Tachiya, 2002; Marcus, 1963; Pope & Swenberg, 1999; Warshel & Parson, 2001) is developed for characterization of such kinds of processes. It describes the environment-induced electron transfer between two adiabatic potentials having the common nuclear configuration coordinate r. The two potentials are described by parabolic shapes, which are shifted with respect to each other by d: $V_i(r) = kr^2/2$, $V_f(r) = k(r-d)^2/2 - \Delta$. Indices i and f denote the initial and final electronic configurations. Here r is the nuclear reorganization coordinate and Δ is the energy gap between the equilibrium states (minima of the potentials). Thus the parameter $kd^2/2 = \lambda$ is the reorganization energy, which characterizes the nuclear relaxation due to electronic transition. The two potentials cross at $V_f(r_t) = V_i(r_t)$; r_t denotes the transfer configuration $r_t = d(\lambda - \Delta)/2\lambda$, where the energies of two states are equal and the transfer is exactly allowed by the energy conservation. The electron transfer happens due to diabatic coupling between these states, J. If J is small, it minimally perturbs the potential surfaces and the transfer rate is given by the probability to reach the crossing point from $r = 0$, which is possible due to thermal activation. Thus, the J coupling is included perturbatively up to the second order and the rate of the electron transfer in this case is given by (Marcus & Sutin, 1985):

$$k_{if}(\Delta, \lambda) = \frac{2\pi}{\hbar^2}|J|^2 \exp\left(-\frac{(\lambda - \Delta)^2}{4\lambda k_B T}\right) \tag{28}$$

This expression holds when the nuclear dynamics is much faster than the transfer time, hence, the thermal equilibrium at the initial state is not perturbed by the presence of the final potential. This covers two regimes of nuclear dynamics. In the overdamped regime the nuclear coordinate shows Brownian fluctuations and the transfer occurs as the system reaches the r_t point. However, in the weakly damped case the nuclei oscillate coherently and can cross the point r_t many times before transfer occurs (the so-called Landau–Zener regime) (May & Kühn, 2011; Weiss, 2008).

When the diabatic coupling parameter J is not vanishing, it may perturb the diabatic potential surfaces and introduce the splitting between new adiabatic potentials. In this case the transfer proceeds on the lower potential surface and thermal activation barrier is reduced. The transfer is thus necessarily mediated by the nuclear dynamics. Extensions of the Marcus theory include the intramolecular vibrational and diffusive solvent orientational motions (Heitele, 1993; Sumi & Marcus, 1986). It was found that the time dependence of the survival probability becomes multiexponential rather than single exponential. Additional modulation of the transfer rates has been discussed in 2010s (Abramavicius, Gulbinas, & Valkunas, 2003; Abramavicius & Valkunas, 2016; Fuller et al., 2014).

One important distinction of the Marcus theory from the Redfield and Förster theories is the explicit presence of the activation energy. Notice that even for the down-hill transfer the activation energy is also invoked in this case and only when the energy gap matches the reorganization energy, the activation barrier disappears. In all other cases the thermal activation is necessary to reach the transfer configuration.

3.4 Exact Description: Hierarchical Equations of Motion for Excitons

Hierarchical equations of motion (HEOM) is a nonperturbative theory describing the exciton dynamics in the open quantum systems (Ishizaki & Fleming, 2009b; Tanimura, 2006). Because the usage of the full theory is computationally expensive a modified HEOM theory, termed as hierarchical quantum master equation (HQME) is often employed (Valkunas et al., 2013; Xu, Tian, Xu, Shi, & Yan, 2009). It is still nonperturbative, but is restricted to the following approximate form of the bath correlation function:

$$C_n(t) = \left(\frac{2\lambda_n}{\beta_{\mathrm{T}}} - \frac{\beta_{\mathrm{T}}\lambda_n\gamma_n^2}{6}\right)\mathrm{e}^{-\gamma_n t} - \iota\lambda_n\gamma_n\mathrm{e}^{-\gamma_n t} + \frac{\lambda_n\gamma_n\beta_{\mathrm{T}}}{3}\delta(t). \tag{29}$$

where $\beta_{\mathrm{T}} = 1/k_{\mathrm{B}}T$ is the inverse temperature.

This form can be obtained from the overdamped Brownian oscillator spectral density, given by

$$C''_n(\omega) = 2\lambda_n\frac{\gamma_n\omega}{\omega^2 + \gamma_n^2} \tag{30}$$

and expanding the Bose–Einstein function up to $(\beta_{\mathrm{T}}\omega)^1$ term:

$$\frac{1}{1 - \mathrm{e}^{-\beta_{\mathrm{T}}\omega}} = \frac{1}{\beta_{\mathrm{T}}\omega} + \frac{1}{2} + \frac{\beta_{\mathrm{T}}\omega}{12} + \mathcal{O}\left[(\beta_{\mathrm{T}}\omega)^3\right]. \tag{31}$$

Conventional high-temperature approximation schemes use only the first two terms of this expansion. However, in general all three terms can be taken into consideration and the rest of the correlation function might be accounted as the Markovian-white-noise residue ansatz (Xu et al., 2009).

The HQME for the correlation function given by Eq. (29) is written in the Liouville space as a hierarchy of coupled differential equations for auxiliary density operators denoted by $|\rho_{\mathbf{n}}(t)\rangle\rangle$ (Xu et al., 2009):

$$\frac{d}{dt}|\rho_{\mathbf{n}}(t)\rangle\rangle = -\iota\widehat{\mathscr{L}}_e|\rho_{\mathbf{n}}(t)\rangle\rangle - \sum_{m=1}^{N}\left(\gamma_m n_m + \delta\widehat{\mathscr{R}}_m\right)|\rho_{\mathbf{n}}(t)\rangle\rangle + \sum_{m=1}^{N} n_m \widehat{\mathscr{A}}_m|\rho_{\mathbf{n}_m^-}(t)\rangle\rangle + \iota\sum_{m=1}^{N}\widehat{\mathscr{Q}}_m^{\times}|\rho_{\mathbf{n}_m^+}(t)\rangle\rangle \tag{32}$$

where the auxiliary superoperators:

$$\delta\widehat{\mathscr{R}}_m = \frac{\lambda_m\gamma_m\beta_{\mathrm{T}}}{3}\widehat{\mathscr{Q}}_m^{\times}\widehat{\mathscr{Q}}_m^{\times}, \tag{33}$$

$$\widehat{\mathscr{A}}_m = \iota\left(\left(\frac{2\lambda_m}{\beta} - \frac{\beta_{\mathrm{T}}\lambda_m\gamma_m^2}{6}\right)\widehat{\mathscr{Q}}_m^{\times} - \iota\lambda_m\gamma_m\widehat{\mathscr{Q}}_m^{\circ}\right) \tag{34}$$

are introduced. Here $\widehat{\mathscr{Q}}_m^{\times}\bullet \Leftrightarrow [\widehat{\mathrm{Q}}_m, \bullet]$ denotes the commutator and $\widehat{\mathscr{Q}}_m^{\circ}\bullet \Leftrightarrow \{\widehat{\mathrm{Q}}_m, \bullet\}$ is the anti-commutator. In Eq. (32) $|\rho_0(t)\rangle\rangle$ corresponds to the physical reduced density operator, $\mathbf{n}$ is a vector of indices $\mathbf{n} \equiv (n_1, n_2, \ldots, n_N)$ and the notation $\mathbf{n}_m^{\pm} \equiv (n_1, n_2, \ldots, n_m \pm 1, \ldots, n_N)$ are usually used. All auxiliary operators with any negative index are set to zero.

Formally the hierarchy in Eq. (32) is infinite, thus the equations are non-perturbative and non-Markovian. For practical purposes various truncations schemes could be used. The simplest one is based on the assumption, that all ADOs with their tier level $L = \sum_{m=1}^{N} n_m$ greater than truncation level L_{trunc} are simply discarded. The truncation level is usually chosen to guarantee the convergence of the simulation results. Other truncation schemes are also possible (Chen, Zheng, Shi, & Yan, 2009).

Since the HEOM theory is derived as an operator equation and makes no approximation for the bath, it is independent of the basis chosen for the problem solution. It can thus capture such effects as the polaron formation and localization of an exciton onto a single chromophore. Therefore, the theory can describe much broader class of problems.

4. STOCHASTIC WAVEFUNCTION APPROACHES

All methods relying on the calculation of the reduced density operator share a common property which allows only the investigation of the statistically averaged behaviour of the system. However, in this case the information about the instantaneous dynamic characteristics of the system is lost. To investigate them the wavefunction description of the system is preferable. One of the approximate approaches for the wavefunction is based on the smart guess of parametrized wavefunction (so-called ansatz). A variational method is then used to determine parameters characterizing the wavefunction (Chorosajev et al., 2014; Scot, 1992; Sun, Luo, & Zhao, 2010) based on the statement that the wavefunction has to approximately satisfy the Schrödinger equation. However, the solution is restricted to the specific domain of the ansatz. Alternative approaches for the wavefunction can be classified as quantum jump methods and quantum state diffusion methods. The quantum jump methods are based on a deterministic evolution of the system wave vector with random jumps of the system state, for example, surface hopping when some vibrational adiabatic coordinate is explicitly included (Beenken, Dahlbom, Kjellberg, & Pullerits, 2002; Dahlbom, Beenken, Sundström, & Pullerits, 2002) and they govern the jump rates, or where jumps are realized by explicit jump operators (so-called quantum Monte-Carlo approach) (Piilo, Härkönen, Maniscalco, & Suominen, 2009; Piilo, Maniscalco, Härkönen, & Suominen, 2008). Quantum state diffusion methods propagate the system wave vector under the influence of continuous fluctuations which represent the action of the environment (Diósi, Gisin, & Strunz, 1998; Gisin & Percival, 1993; Yu, Diósi, Gisin, & Strunz, 1999).

4.1 Semiclassical Polaron Model

Poralon formation is a general feature of electronic excitation due to their interaction with molecular vibrations and/or phonons. Instead of second quantization here we use the coordinate/momentum notation:

$$\widehat{H}_{\text{ph}} = \frac{1}{2} M \sum_{\alpha} \left(\dot{\widehat{q}}_{\alpha}^{2} + \omega_{\alpha}^{2} \widehat{q}_{\alpha}^{2} \right), \tag{35}$$

where M is the effective mass of the environmental oscillator. We next associate a single coordinate (vibronic mode) to a specific molecule and assume that all sites are equivalent. Then the interaction Hamiltonian is

$$\widehat{H}_{\text{int}} = -\chi \sum_{n} \widehat{q}_{n} \widehat{B}_{n}^{\dagger} \widehat{B}_{n}, \tag{36}$$

χ is the interaction strength parameter.

Since we are interested in the electronic system properties, we can take the trace of the whole Hamiltonian with respect to the canonical ensemble of bath oscillators; the trace of the coordinate operator becomes the coordinate expectation value. In this case we have for the *n*th site's bath and system-bath interaction:

$$\widehat{H}_n = \frac{1}{2}M\left(\dot{q}_n^2 + \omega_0^2 q_n^2\right) - \chi q_n Tr\left\{\widehat{\rho}\widehat{B}_n^\dagger \widehat{B}_n\right\}, \tag{37}$$

and $Tr\left\{\widehat{\rho}\widehat{B}_n^\dagger \widehat{B}_n\right\} \equiv \rho_{nn}$, thus finally we have

$$\widehat{H} = \frac{1}{2}M\sum_n \left(\dot{q}_n^2 + \omega_0^2 q_n^2\right) - \chi q_n \rho_{nn}. \tag{38}$$

Here q_n is the expectation value of the coordinate operator of the *n*th oscillator and ω_0 is the ground-state vibrational frequency.

Microscopic equations of motion can be obtained by taking a single member of the statistical ensemble (assuming that it is possible to make such separation), which constitutes the pure system, and hence we have $\rho_{mn} = \phi_m(t)\phi_n^*(t)$ from the Hamilton equations which yields (Bang, Christiansen, If, Rasmussen, & Gaididei, 1994)

$$M\ddot{q}_n + M\lambda\dot{q}_n + M\omega_0^2 x_n = \chi|\phi_n(t)|^2 + \eta_n(t) \tag{39}$$

and

$$i\dot{\phi}_n = \varepsilon_n\phi_n + \sum_m^{m\neq n} J_{mn}\phi_m + \chi q_n(t)\phi_n. \tag{40}$$

In Eqs (39) and (40) the term η_n represents the white noise induced by the *remaining* Brownian modes of the environment, which are being reduced into white noise model. The phonon evolution is accompanied by the damping term λ due to the fluctuation–dissipation theorem (Valkunas et al., 2013).

The solution for phonon coordinates can be obtained analytically since their equations are linear in q:

$$\begin{aligned} q_n(t) &= \mathrm{e}^{-\lambda t/2}\left(c_{1n}\cos(\omega_1 t) + c_{2n}\sin(\omega_1 t)\right) \\ &\quad + \frac{1}{M\omega_0}\int_0^t d\tau \sin(\omega_1(t-\tau))\mathrm{e}^{-\lambda(t-\tau)/2}\left(\chi|\phi_n(\tau)|^2 + \eta_n(\tau)\right) \end{aligned} \tag{41}$$

Constants c_1 and c_2 reflect the initial condition, $\omega_1 = \sqrt{\omega_0^2 - (\lambda/2)^2}$.

By considering high frequency molecular vibrations, the oscillator coordinate of which varies much faster than the electronic population, we have

$$
\begin{aligned}
q_n(t) &= \mathrm{e}^{-\lambda t/2}(c_{1n}\cos(\omega_1 t) + c_{2n}\sin(\omega_1 t)) \\
&\quad + \frac{\chi|\phi_n(\tau)|^2}{M\omega_0}\int_0^t d\tau \sin(\omega_1(t-\tau))\mathrm{e}^{-\lambda(t-\tau)/2} \\
&\quad + \frac{1}{M\omega_0}\int_0^t d\tau \sin(\omega_1(t-\tau))\mathrm{e}^{-\lambda(t-\tau)/2}\eta_n(\tau).
\end{aligned}
\tag{42}
$$

And in the case of the white noise spectrum with random phase $\delta(\omega)$, thus, giving

$$
\eta_n(t) = z\int \frac{d\omega}{2\pi}\cos(\omega t + \delta_n(\omega)) \tag{43}
$$

at $t > \lambda^{-1}$, $\omega_1 \gg \lambda$ we have

$$
q_n(t) = \frac{\chi|\phi_n(t)|^2\omega_1}{M\omega_0\left(\omega_1^2 + (\lambda/2)^2\right)} + \frac{1}{M\omega_0}s_n(t),
$$

where

$$
s_n(t) = z\int \frac{d\omega}{2\pi}\frac{\omega_1}{\left(\omega_1^2 - \omega^2\right)^2 + \omega^2\lambda^2}\cos(\omega t + \delta_n(\omega)) \tag{44}
$$

is the coloured noise filtered through molecular vibrational resonance. This description corresponds the situation when the coordinate follows the excitonic dynamics adiabatically generating the coloured noise.

The resulting equations for electronic amplitudes become nonlinear signifying polaronic special solutions.

4.2 Variational Wavefunction Approach

A set of equations similar to the semiclassical approach can be obtained by using more rigorous Dirac variational principle. To follow the excitation dynamics in a molecular system let us consider the single-excitation subspace with the ground states of the exciton and phonon subsystems correspondingly denoted by $|0\rangle_{\mathrm{ex}}$, $|0\rangle_{\mathrm{ph}}$ and the global ground state by $|\varnothing\rangle$. Instead of solving the Schrödinger equation let us employ the Dirac—Frenkel variational principle to study the excitation dynamics (Frenkel & Frenkel, 1936). In this case we use a parametrized trial wavefunction and optimize the variational parameters so that the deviation from the solution of the Schrödinger equation would be minimal. The Davydov D_1 ansatz is chosen as the trial wavefunction (Chorosajev et al., 2014; Huynh, Sun, Gelin, & Zhao, 2013; Scot, 1992; Sun et al., 2010; Ye et al., 2012):

$$|\Psi_{D1}(t)\rangle = \sum_{n}\left(\alpha_n(t)\hat{a}_n^{\dagger}\exp\left[\sum_{q}\lambda_{qn}(t)\hat{b}_q^{\dagger} - \text{h.c.}\right]\right)|\varnothing\rangle. \tag{45}$$

In this form the wavefunction of the system is a superposition of complex electronic excitation amplitudes $\alpha_n(t)$ multiplied by coherent states for the phonon field modes, displaced by $\lambda_{qn}(t)$. Hence, the whole Hilbert space for the solutions is contracted to a subspace spanned by the parametrized wavefunction and the minimal uncertainty states for the phonon field are admitted. Considering N sites and Q modes for the phonon field, the wavefunction then requires $N(Q+1)$ complex time-dependent parameters to be uniquely specified.

The procedure to obtain the parameter dependences on time that would minimize the deviation from the exact solution is as follows. First, we construct the Lagrangian, which is given by

$$\mathscr{L} = \frac{i}{2}\left(\langle\Psi_{D1}(t)|\dot{\Psi}_{D1}(t)\rangle - \langle\dot{\Psi}_{D1}(t)|\Psi_{D1}(t)\rangle\right) - \langle\Psi_{D1}(t)|\hat{H}|\Psi_{D1}(t)\rangle. \tag{46}$$

With the Lagrangian we derive the equations of motion (EOM) using the Euler–Lagrange equations for the variational parameters $\{\alpha_n^*(t), \lambda_{qn}^*(t)\}$:

$$\frac{d}{dt}\left(\frac{\partial\mathscr{L}}{\partial\dot{\alpha}_n^*}\right) - \frac{\partial\mathscr{L}}{\partial\alpha_n^*} = 0, \tag{47}$$

$$\frac{d}{dt}\left(\frac{\partial\mathscr{L}}{\partial\dot{\lambda}_{qn}^*}\right) - \frac{\partial\mathscr{L}}{\partial\lambda_{qn}^*} = 0, \tag{48}$$

The resulting equations describing the time dependence of parameters are as follows (Chorosajev et al., 2014):

$$\begin{aligned}\dot{\alpha}_n = &-\frac{i}{2}\sum_{m}\left(J_{nm}\alpha_m S_{n,m}\sum_{q}\left[|\lambda_{qn}|^2 - \lambda_{qn}^*\lambda_{qm}\right]\right)\\ &-\frac{i}{2}\sum_{m}\left(J_{nm}\frac{\alpha_m^*\alpha_n}{\alpha_n^*}S_{n,m}^*\sum_{q}\left[|\lambda_{qn}|^2 - \lambda_{qn}\lambda_{qm}^*\right]\right)\\ &-i\sum_{m}J_{nm}\alpha_m S_{n,m} + \frac{i}{2}\alpha_n\sum_{q}g_{qn}\omega_q\left(\lambda_{qn} + \lambda_{qn}^*\right),\end{aligned} \tag{49}$$

$$\dot{\lambda}_{qn} = i\sum_m J_{nm}\frac{\alpha_n^*\alpha_m}{|\alpha_n|^2}S_{n,m}\left(\lambda_{qm} - \lambda_{qn}\right) - i\omega_q\lambda_{qn} + i\omega_q g_{qn}. \tag{50}$$

Here $S_{n,m}$ is the Debye—Waller factor determining the overlap of two coherent states corresponding to sites n and m. It is given by

$$S_{n,m} = \exp\sum_q\left(\lambda_{qn}^*\lambda_{qm} - \frac{1}{2}\left[|\lambda_{qn}|^2 + |\lambda_{qm}|^2\right]\right). \tag{51}$$

Properties of the phonon field can be represented by a spectral density function, which in the case of the set of discrete modes is given by (Valkunas et al., 2013):

$$C''_{mn}(\omega) = \pi\sum_q g_{qm}g_{qn}\omega_q^2\delta\left(\omega - \omega_q\right). \tag{52}$$

The coupling strength between the electronic and phonon coordinates is now characterized by a reorganization energy matrix

$$\Lambda_{mn} = \frac{1}{\pi}\int_0^\infty \frac{C''_{mn}(\omega)}{\omega}d\omega = \sum_q g_{qm}g_{qn}\omega_q. \tag{53}$$

For molecular systems the phonon modes are mostly localized. For specific calculations it is usually assumed that each phonon mode is coupled to a single site, and thus fluctuations of site energies are independent and the same for all sites. The spectral density then reduces to a single function, $C''_{mn}(\omega) = \delta_{mn}C''(\omega)$. To take into account the dissipative nature of the environmental dynamics the spectral density is very frequently represented as the function corresponding to an overdamped oscillator (Butkus et al., 2012).

This approach essentially results in an initial value problem that can be solved using common methods. Again the equations of motion are nonlinear and thus imply the specific polaronic states.

4.3 Perturbational Stochastic Equation

In this section we describe the case where the stochastic equation for the wavefunction will be derived from the Schrödinger equation by postulating the weak system—bath interaction (Abramavicius & Abramavicius, 2014).

As it was outlined in Section 2 of this chapter an arbitrary molecular system can be described by the system Hamiltonian $\hat{H}_S$ and its environment — the bath with the Hamiltonian $\hat{H}_B$. In this section we assume that the system is linearly coupled to the environment via a set of system operators $\widehat{L}_n$ and $\widehat{L}_n^\dagger$, thus the system—bath interaction Hamiltonian will be written as

$$\widehat{H}_{\mathrm{SB}} = \kappa \sum_n \sum_j \left[\widehat{L}_n g_{nj} \widehat{a}_j^\dagger + \mathrm{h.c.} \right]. \tag{54}$$

Here the quantity κ parametrizes the overall strength of the interaction between the system and the environment; g_{nj} and g_{nj}^* are constants describing the coupling strength between the jth bath oscillator and the nth system operator $\widehat{L}_n$.

The composite system is closed; thus its state can be described by the wave vector $|\Psi(t)\rangle$ which should satisfy the Schrödinger equation

$$i\frac{d}{dt}|\Psi(t)\rangle = \left(\widehat{H}_{\mathrm{S}} + \widehat{H}_{\mathrm{B}} + \widehat{H}_{\mathrm{SB}}\right)|\Psi(t)\rangle \equiv \widehat{H}|\Psi(t)\rangle. \tag{55}$$

The solution of this equation formally can be written using the evolution operator $\widehat{U}_0(t)$:

$$|\Psi(t)\rangle = \widehat{U}_0(t)|\Psi(0)\rangle = \mathrm{e}^{-i\widehat{H}t}|\Psi(0)\rangle. \tag{56}$$

Let us now switch to the interaction representation with respect to the bath Hamiltonian $\widehat{H}_{\mathrm{B}}$. In this representation we find a new time-dependent Hamiltonian of the composite system as

$$\mathrm{e}^{i\widehat{H}_{\mathrm{B}}t}\left(\widehat{H}_{\mathrm{S}} + \widehat{H}_{\mathrm{SB}}\right)\mathrm{e}^{-i\widehat{H}_{\mathrm{B}}t} = \widehat{H}_{\mathrm{S}} + \kappa \sum_n \sum_j \left[\widehat{L}_n g_{nj} \widehat{a}_j^\dagger \mathrm{e}^{i\omega_j t} + \mathrm{h.c.} \right]. \tag{57}$$

The wave vector, in the interaction representation $|\Psi(t)\rangle \to |\Psi(t)\rangle^{(I)} = \mathrm{e}^{i\widehat{H}_{\mathrm{B}}t}|\Psi(t)\rangle$, satisfies the Schrödinger Eq. (55) with the transformed Hamiltonian. Its solution is formally given by

$$|\Psi(t)\rangle^{(I)} = \widehat{U}(t)|\Psi(0)\rangle. \tag{58}$$

Further on in this section we work only in the interaction representation, hence for brevity we drop index (I) above the wave vector. The wave vector $|\Psi(t)\rangle$ of the composite system encodes the full information about the evolution of both the environment and the system. However, we are only interested in the dynamics of the latter.

The time-dependent Schrödinger equation implies the initial value problem for the wavefunction. Let us consider the initial conditions at finite temperature. We assume that initially the interaction between the system and the environment is turned off (which is the case when the system is in the ground state) and the system is not correlated with the environment. As the excitation timescale is much shorter than nuclear dynamics, the state of the system just after excitation is still uncorrelated with the environment and can be defined as pure excitonic state $|\phi\rangle$. For the bath we must have the thermal equilibrium, which is described by the density operator. Hence, the initial condition can be defined for the density operator of the composite system as a tensor product (Diosi & Strunz, 1997):

$$\hat{\rho}(0) = |\phi\rangle\langle\phi| \otimes Z^{-1} \mathrm{e}^{-\beta_{\mathrm{T}} \hat{H}_{\mathrm{B}}}, \tag{59}$$

Here $Z = Tr_{\mathrm{B}}[\mathrm{e}^{-\beta_{\mathrm{T}}\hat{H}_{\mathrm{B}}}]$, while the latter trace is over the bath degrees of freedom.

As all bath oscillators and the system are uncoupled at $t = 0$, the wave vector $|\Psi(0)\rangle$ of the composite system can be formally written in a chosen basis for the bath states $|\beta_j\rangle$ as a tensor product as well

$$|\Psi(0)\rangle = |\phi\rangle \otimes |\beta_1\rangle \otimes |\beta_2\rangle \otimes \ldots \equiv |\phi\rangle \otimes |\boldsymbol{\beta}\rangle, \tag{60}$$

It is convenient to describe the state of the environmental oscillators using the coherent states of harmonic oscillators (see eg, Gazeau, 2009; Valkunas et al., 2013; Zhang, Feng and Gilmore, 1990, for details). By taking the dot product of the full-wave vector $|\Psi(0)\rangle$ and the conjugate vector $\langle\boldsymbol{\alpha}| = \langle\alpha_1| \otimes \langle\alpha_2| \otimes \ldots$, we define the wave vector of the system at the initial time with respect to the chosen right-side bath configuration as

$$\left|\psi\boldsymbol{\alpha}_{\phi\boldsymbol{\beta}}(0)\right\rangle = \mathrm{e}^{-\boldsymbol{\alpha}^*\boldsymbol{\beta}}\langle\boldsymbol{\alpha}|\Psi(0)\rangle, \tag{61}$$

where $\boldsymbol{\alpha}^*\boldsymbol{\beta} = \sum_{j=1}^{\infty} \alpha_j^* \beta_j$ are due to nonorthogonality of coherent states. At the time t we can then write

$$\left|\psi\boldsymbol{\alpha}_{\phi\boldsymbol{\beta}}(t)\right\rangle = \mathrm{e}^{-\boldsymbol{\alpha}^*\boldsymbol{\beta}}\langle\boldsymbol{\alpha}|\Psi(t)\rangle \equiv \mathrm{e}^{-\boldsymbol{\alpha}^*\boldsymbol{\beta}}\left\langle\boldsymbol{\alpha}\right|\hat{U}(t)\left|\boldsymbol{\beta}\right\rangle|\phi\rangle. \tag{62}$$

Recall that $|\phi\rangle$ does not depend on the bath state, hence, it has no indices α and β.

The system wave vector $\left|\psi\boldsymbol{\alpha}_{\phi\boldsymbol{\beta}}(t)\right\rangle$ is a stochastic system wave vector depending on the particular configuration of the environment, characterized by two complex-valued infinite-dimensional vectors $\boldsymbol{\alpha}$ and $\boldsymbol{\beta}$. Consequently, we can take one particular configuration ($\boldsymbol{\alpha}$,$\boldsymbol{\beta}$), calculate the vector $\left|\psi\boldsymbol{\alpha}_{\phi\boldsymbol{\beta}}(t)\right\rangle$ and perform the ensemble averaging with the probability density (Abramavicius & Abramavicius, 2014)

$$p_{\alpha\beta}(\omega_j) = \overline{n}^{-1}(\omega_j)\exp\left(-|\alpha|^2 - |\beta|^2 \mathrm{e}^{\beta_{\mathrm{T}}\omega_j} + \alpha^*\beta + \alpha\beta^*\right). \tag{63}$$

where $\overline{n}(\omega) = (\exp(\beta_T\omega) - 1)^{-1}$ is the Bose-Einstein function.

The averaging leads to the proper reduced density matrix and all other proper statistical parameters. The vectors $\boldsymbol{\alpha}$ and $\boldsymbol{\beta}$ can be treated as complex-valued random vectors.

The equation for the system wave vector $|\psi\boldsymbol{\alpha}_{\phi\boldsymbol{\beta}}(t)\rangle$ can be obtained by differentiating Eq. (62) with respect to time:

$$\begin{aligned}\frac{d}{dt}|\psi\boldsymbol{\alpha}_{\phi\boldsymbol{\beta}}(t)\rangle &= \mathrm{e}^{-\boldsymbol{\alpha}^*\boldsymbol{\beta}}\langle\boldsymbol{\alpha}|\frac{d}{dt}\widehat{U}(t)|\boldsymbol{\beta}\rangle|\phi\rangle \\ &= -i\widehat{H}_{\mathrm{S}}|\psi\boldsymbol{\alpha}_{\phi\boldsymbol{\beta}}(t)\rangle - i\kappa\sum_{nj}\widehat{L}_n g_{nj}\mathrm{e}^{i\omega_j t}\alpha_j^*|\psi\boldsymbol{\alpha}_{\phi\boldsymbol{\beta}}(t)\rangle \\ &\quad - i\mathrm{e}^{-\boldsymbol{\alpha}^*\boldsymbol{\beta}}\kappa\sum_{nj}\widehat{L}_n^\dagger g_{nj}^*\mathrm{e}^{-i\omega_j t}\langle\boldsymbol{\alpha}|\widehat{a}_j\widehat{U}(t)|\boldsymbol{\beta}\rangle|\phi\rangle.\end{aligned} \tag{64}$$

In the last term of Eq. (64) the quantity $\langle\boldsymbol{\alpha}|\widehat{a}_j\widehat{U}(t)|\boldsymbol{\beta}\rangle|\phi\rangle$ can be evaluated at first order in κ. The resulting stochastic Schrödinger equation is closed and time local:

$$\begin{aligned}i\frac{d}{dt}|\psi(t)\rangle &= \widehat{H}_{\mathrm{S}}|\psi(t)\rangle + \kappa\sum_n\left[\widehat{L}_n z_n(t) + \widehat{L}_n^\dagger w_n(t)\right]|\psi(t)\rangle \\ &\quad - i\kappa^2\sum_{nm}\left[\widehat{L}_n^\dagger\int_0^t d\tau C_{nm}^{(0)}(\tau)\mathrm{e}^{-i\widehat{H}_{\mathrm{S}}\tau}\widehat{L}_m\mathrm{e}^{i\widehat{H}_{\mathrm{S}}\tau}\right]|\psi(t)\rangle,\end{aligned} \tag{65}$$

where we dropped indices α and β ($|\psi\boldsymbol{\alpha}_{\phi\boldsymbol{\beta}}(t)\rangle \equiv |\psi(t)\rangle$) as they do not enter explicitly. Additionally, we defined the following quantities with respect to the bath oscillators:

$$z_n(t) = \sum_j g_{nj}\alpha_j^*\mathrm{e}^{i\omega_j t}, \tag{66}$$

$$w_n(t) = \sum_j g_{nj}^*\beta_j\mathrm{e}^{-i\omega_j t}, \tag{67}$$

$$C_{nm}^{(0)}(t) = \sum_j g_{nj}^* g_{mj}\mathrm{e}^{-i\omega_j t}. \tag{68}$$

Since α_j and β_j are random, $z_n(t)$ and $w(t)$ are Fourier transformations of these from the frequency domain to the time domain. This means that $z_n(t)$ and $w_n(t)$ are complex-valued fluctuations. By calculating their correlation functions $\mathscr{Z}_{nn}(t) = \langle z_n^*(t)z_n(0)\rangle_{\mathrm{ens}}$ and $\mathscr{W}_{nn}(t) = \langle w_n^*(t)w_n(0)\rangle_{\mathrm{ens}}$, where

$\langle\ldots\rangle_{\text{ens}}$ denotes the statistical averaging operation using the Gaussian probability density function from Eq. (63), we find that

$$\mathscr{Z}_{nm}(t) = \sum_j \left(\bar{n}(\omega_j) + 1\right) g_{nj}^* g_{mj} e^{-i\omega_j t} \tag{69}$$

and for $w_n(t)$,

$$\mathscr{W}_{nm}(t) = \sum_j \bar{n}(\omega_j) g_{nj} g_{mj}^* e^{i\omega_j t}. \tag{70}$$

These functions depend on temperature, however, as $T \to 0$ we have $C_{nm}^{(0)}(t) \equiv \mathscr{Z}_{nm}(t)\big|_{T=0}$.

Next, each molecule is usually considered as a two-level system, characterized by the excitation energy ε_n. Let us consider only a single excitation in the aggregate, thus suggesting that state $|n\rangle$ describes the situation with excitation residing on site n. It is often assumed that the interaction of such system with the environment can be approximated by including the diagonal fluctuations (to excitation energies). For diagonal-only fluctuations, $\hat{L}_n = |n\rangle\langle n|$. Thus, the fluctuations of the heat bath affect only the diagonal elements of the system Hamiltonian $\hat{H}_S$. Additionally, we assume that different projectors $\hat{L}_n$ are coupled to different sets of the bath oscillators. This makes the correlation functions $\mathscr{Z}_{nm}(t)$ and $\mathscr{W}_{nm}(t)$ diagonal. Taking that the environment of all sites is statistically the same $\left(C_{nm}^{(0)}(\tau) \equiv \delta_{nm} C^{(0)}(\tau)\right)$ Eq. (65) for the system wave vector $|\psi(t)\rangle$ can then be written in the following way:

$$\begin{aligned} i\frac{d}{dt}|\psi(t)\rangle &= \hat{H}_S|\psi(t)\rangle + \kappa \sum_n \hat{L}_n u_n(t)|\psi(t)\rangle \\ &\quad - i\kappa^2 \sum_n \left[\hat{L}_n \int_0^t d\tau C^{(0)}(\tau) e^{-i\hat{H}_S\tau} \hat{L}_n e^{i\hat{H}_S\tau}\right] |\psi(t)\rangle \end{aligned} \tag{71}$$

where we have a new stochastic function $u_n(t) = z_n(t) + w_n(t)$. The stochastic function $u_n(t)$ replaces functions $z_n(t)$ and $w_n(t)$. Hence, the set of variables $(\boldsymbol{\alpha},\boldsymbol{\beta})$ can now be replaced by stochastic complex-value functions of frequency $u_n(\omega)$. The sole characteristics which fully defines $u_n(t)$ and $u_n(\omega)$ is the correlation function of u. Using Eqs (69) and (70) we find

$$C_{nn}(t) = \sum_j \left[\left(\bar{n}(\omega_j) + 1\right)\left|g_{nj}\right|^2 e^{-i\omega_j t} + \bar{n}(\omega_j)\left|g_{nj}\right|^2 e^{i\omega_j t}\right]. \tag{72}$$

The correlation function can be fully defined by the spectral density, which is a continuous function of frequency for an infinite number of bath oscillators:

$$C_{nn}(t) \equiv C(t) = \int \frac{d\omega}{2\pi} C''(\omega) \times \left(\coth \frac{\omega \beta_T}{2} \cos\omega t - i\sin\omega t \right). \tag{73}$$

Additionally, since $\mathscr{W}_{nm}(t) = 0$ at zero temperature, we have

$$C_{nn}^{(0)}(\tau) = \int \frac{d\omega}{2\pi} C''(\omega) e^{-i\omega t}. \tag{74}$$

That is, it is a Fourier transform of the spectral density.

5. ELECTRON TRANSFER IN A MESOSCOPIC MOLECULAR NETWORK

Approaches that are described in the previous section are usually applied for small complexes of few molecules or sites. However the artificial photosynthetic systems can be composed out of a large number of constituent electron sites. Such systems are, for instance, the polymers or their blends, which are extensively studied for photovoltaics.

A major improvement in solar cell efficiency was obtained by the introduction of the bulk heterojunction (BHJ) device structure (Halls et al., 1995; Yu, Gao, Hummelen, Wudl, & Heeger, 1995) consisting of a mesostructured blend film of conjugated polymer and fullerene. Upon light absorption, excitons are produced in the polymer phase and are converted into closely bound electron—hole pairs at the polymer—fullerene interface, sometimes also referred to as charge transfer (CT) states. The bound charge pair states have to dissociate in order to produce photocurrent by overcoming the mutual Coulomb attraction between electrons and holes into the free mobile charges. It has been argued that only the closely separated charge pair states are initially created (van Eersel, Janssen, & Kemerink, 2012; Lee et al., 2010; Marsh, Hodgkiss, & Friend, 2010) and favourable structural organization is more important than excess energy (Herrmann et al., 2011; Keivanidis et al., 2010; Vithanage et al., 2013].

The electron motion in the large network of sites, which reflect either polymer, or fullerene or both, can be realized by the Monte-Carlo simulation protocol with predefined electron transport rates (Bässler, 1993). In the whole system the electron and hole dynamics are controlled by site-energy landscape. Consider one charge (electron) mobile in a lattice. In the presence of an external electric field the energy of the electron in the lattice consists of three parts: (1) the internal site self-energy E_r, which is constant on average, however, due to inhomogeneities it is usually assumed to be a random Gaussian value, which can be spatially correlated (Abramavicius & Valkunas, 2003), (2) the energy due to the constant external electric field $\boldsymbol{F}$ and (3) the energy due to the Coulomb interaction between opposite charges, $E_C(|\boldsymbol{r}|)$. The electron energy thus equals to

$$E_e(\boldsymbol{r}) = E_r + (\boldsymbol{F}\cdot\boldsymbol{r}) + E_C(|\boldsymbol{r}|) \tag{75}$$

The energy of the external electric field can be accounted for by projecting the site position to the electric field direction. The electrostatic interaction energy is given by the shifted Coulomb potential:

$$E_C(|\boldsymbol{r}|) = \frac{1}{4\pi\varepsilon\varepsilon_0}\cdot\frac{q_e}{|\boldsymbol{r}| + \eta} \tag{76}$$

Here q_e is the electron charge, $|\boldsymbol{r}|$ is the distance from the origin, where the opposite charge is located, ε is the mean permittivity of the material and η is a positive displacement parameter, which accounts for deviation of the Coulomb potential from the point charge approximation at short distances and sets the appropriate initial electron—hole interaction energy.

The Monte-Carlo algorithm is realized as follows. As the initial configuration, the hole and electron are placed on neighbouring sites in the interfacial region of the donor and acceptor domains. A single hopping event is simulated at a time. Only the nearest neighbour sites are taken into account for the electron hopping event. In the cubic lattice the electron can hop into one of six surrounding sites when it is far from the interface while hopping possibilities are fewer in the interfacial region.

Consider hopping originating from site m. The pool of final (target) sites T is created. The hopping rate from site m into $n \in T$, ν_{mn} can be calculated using the Marcus formula or the Miller—Abrahams formula:

$$\nu_{mn} = \nu_0 \exp(-2\gamma|\boldsymbol{r}_m - \boldsymbol{r}_n|) \times \begin{cases} \exp\left(-\dfrac{E_m - E_n}{k_B T}\right) & , E_m < E_n \\ 1 & , E_m > E_n \end{cases} \tag{77}$$

where γ is a parameter which characterizes the inverse localization length of a charge density, and $E_m \equiv E_e(\boldsymbol{r}_m)$ and $E_n \equiv E_e(\boldsymbol{r}_n)$ are their total energies, respectively.

When all rates of possible hopping events from originating site m have been evaluated, the rates are being translated into hopping probabilities according to

$$p_{mn} = \frac{\nu_{mn}}{\sum_k \nu_{mk}}. \tag{78}$$

These probabilities are then used to determine the destination site for the electron, chosen by a uniformly distributed random number. The charge configuration is then switched to the one that has been determined and the rates of the next hopping events are recalculated.

Up to few thousand electron trajectories can be easily generated and this yields the probability distribution flow, equivalent to the population evolution of the reduced density matrix (Abramavicius & Valkunas, 2003, Vithanage et al., 2013).

6. EXAMPLE: EXCITATION DYNAMICS IN A HETERODIMER

Let us consider a model system, a heterodimer. Usually, a homodimer is used for considering various fundamental aspects of the nuclei in the exciton dynamics and relaxation (Cho & Fleming, 2005; Gelzinis et al., 2011; Kjellberg, Brüggemann, & onu Pullerits, 2006; Pisliakov, Mancal, & Fleming, 2006). For a heterodimer, the distinctness of constituent monomers is often limited to excitation energies (Ishizaki & Fleming, 2009a; Schlau-Cohen, Ishizaki, & Fleming, 2011). Some aspects which could be attributed to the heterodimer were also disclosed by analysing the exciton–CT-state mixing problems (Mancal, Valkunas, & Fleming, 2006; Renger, 2004). Experiments since 2000 based on coherent photon echo measurements have demonstrated the possibility to follow the coherent phase dynamics and incoherent population relaxation of excitons in photosynthetic pigment–protein complexes (van Grondelle & Novoderezhkin, 2006; Heijs, Dijkstra, & Knoester, 2007; Schlau-Cohen et al., 2011]. Possible role of the coherent dynamics in determining such type of the regulation function – attributing it to the ability of the excitation quenching in heterodimers – was considered since mid-2010s (Balevičius et al., 2012, 2013).

The single ground state and the two singly excited states of the dimer are given as the product of the ground states of the monomers (sites) a,b, $|g\rangle = |g_a\rangle|g_b\rangle$, and the product of the ith monomer in the excited state $|e_i\rangle$

and jth monomer in ground states, that is, $|i\rangle = |e_i\rangle|g_j\rangle$, respectively. In this case the Hamiltonian determining the exciton states for the dimer is given by Eqs (3) and (4). All characteristics of the monomers within the dimer discussed in the preceding paragraphs are depicted in Fig. 1.

We switch to the eigenstates of the excitonic system, that is, to the exciton representation, by a unitary transformation that diagonalizes the system Hamiltonian. The obtained excitonic basis is related to the molecular representation accordingly:

$$\begin{cases} |1\rangle = & \cos\theta|a\rangle + \sin\theta|b\rangle; \\ |2\rangle = & -\sin\theta|a\rangle + \cos\theta|b\rangle. \end{cases} \tag{79}$$

The mixing angle $\theta = \frac{1}{2}\arctan(2J/\Delta\varepsilon)(\theta \in [-\pi/4; \pi/4])$ depends on the resonance interaction and the energy gap $\Delta\varepsilon = (\varepsilon_a^0 + \lambda_a) - (\varepsilon_b^0 + \lambda_b)$ between the transition energies of the monomers. This angle describes the degree of delocalization of excitation over the monomeric states.

The dynamics of an open quantum system is given by a reduced density operator (RDO) $\rho(t) = Tr_q W(t)$, where $W(t)$ is the total-density operator of the system and the bath, and Tr_q denotes the trace operation over the nuclear

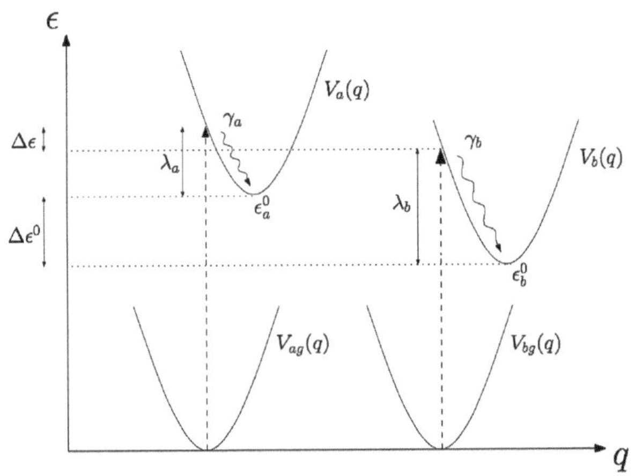

Figure 1 Basic parameters of the model heterodimer. The constituent monomers are depicted as pairs of shifted harmonic oscillators. The *dashed arrows* indicate the Franck–Condon transitions. In the case of isolated monomers an optical excitation is followed by the nuclear reorganization on the excited-state potential energy surface $V_i(q)$, which leads to the dissipation of the reorganization energy λ_i on the time scale of γ_i^{-1} (*wiggly arrows*). We denote the energies of the relaxed excited states by ε^0 to distinguish them from the vertical transition energies, $\varepsilon = \varepsilon^0 + \lambda$. Clearly $\lambda_a \neq \lambda_b$ corresponds to $\Delta\varepsilon \neq \Delta\varepsilon_0$.

degrees of freedom (Butkus, Valkunas, & Abramavicius, 2011; May & Kühn, 2011; Mukamel, 1995; Valkunas et al., 2013; Weiss, 2008). The evolution of the RDO is governed by the Liouville equation:

$$\frac{d}{dt}\rho(t) = -i\mathscr{L}\rho(t) + \mathscr{D}\rho(t) + \mathscr{K}\rho(t). \tag{80}$$

The three superoperators on the RHS represent the unitary evolution, the dissipative dynamics and relaxation to the ground state (RGS), respectively. The Liouvillian defined as $\mathscr{L}\bullet \equiv \frac{1}{\hbar}[H_S, \bullet]$, where the square brackets denote the commutator, generates the quantum coherent evolution. The second term, $\mathscr{D}\rho(t)$ describes the dissipation and pure dephasing within the excited states. The last term is defined as $\mathscr{K}\bullet \equiv -\sum \frac{\kappa_i}{2}\{Q_i, \bullet\}$; with $Q_i = |i\rangle\ \langle i|$, where the curly brackets denote the anticommutator, and κ_i is the decay rate of the *i*th excited state. Parameters κ_i can be phenomenologically defined in the site basis (Balevičius et al., 2012). The form of the RGS superoperator guarantees that the diagonal elements of the RDO (the populations) $\rho_{ii}(t)$ decay with the rates κ_i, while the corresponding off-diagonal elements (the coherences) $\rho_{ij}(t)$ decay with rates $\frac{1}{2}(\kappa_i + \kappa_j)$.

Let us assume now that initially the full-density operator is of the factorized form: $W(0) = \rho(0) \otimes W_{eq}$, where $W_{eq} = e^{-\beta_T H_B} / Tr_q e^{-\beta_T H_B}$ is the equilibrium density operator of the bath. This is an important assumption since it disregards the initial correlations between the system and the bath, however, it is justified in the description of electronic excitations due to the Franck–Condon principle. To be more specific, we further consider the excitation of the system by an ultrashort pulse, which despite its spectral broadness, excites only the monomer *a* while the monomer *b* is assumed to be optically dark (cf. the Car S_1 state). This results in the initial condition in the site basis as $\langle i|\rho(0)|j\rangle = \delta_{ij}\delta_{ia}$.

We set up a case study of a heterodimer which has some typical properties of photosynthetic Chl–Car aggregates. In the following we consider the dynamic effects originated from the differences in the excitation energies, reorganization energies and excitation lifetimes of the constituent molecules of the heterodimer. One of the constituent monomers is characterized by an optically dark and extremely short-lived excited state. The other monomer has a strong transition dipole moment and a long-lived excited state. Their homogeneous broadenings are different as well. This situation is schematically depicted in Fig. 1. The chosen parameters are the following: $T = 300\text{K}$ (which corresponds to $k_B T = 208\ \text{cm}^{-1}$), $J_{ab} = 100\ \text{cm}^{-1}$, $\gamma^{-1} = 100$ fs, dipole moments $d_a = 1$, $d_b = 0$. The tunable parameters are the reorganization energies λ_a, λ_b, and the gap between the site excitation energies. We consider the excitation of the dimer by a broadband optical pulse which is much shorter than the time scale of the relaxation dynamics. Since only the optical excitation of monomer a is allowed the initial condition for the evolution of the reduced density matrix in the site basis is given by $\rho_{aa}(0) = |d_a|^2 = 1$, while other elements are zero. The main parameter characterizing a heterodimer is the difference of site energies.

We study the dynamics of the excited state in the absence of relaxation to the ground state, that is, when $\mathscr{K} = 0$. In Fig. 2 we show the time evolution of the higher excitonic state population modelled by the Redfield equations (a and b) and the HQME (c and d) using four combinations of the reorganization energies. The left column corresponds to $\Delta\varepsilon = +100\ \mathrm{cm}^{-1}$ as a fixed energy gap, while the right column corresponds to $\Delta\varepsilon_0 = +100\ \mathrm{cm}^{-1}$ being fixed. The equal reorganization energies (black and blue curves) serve as a good starting point for comparison of the

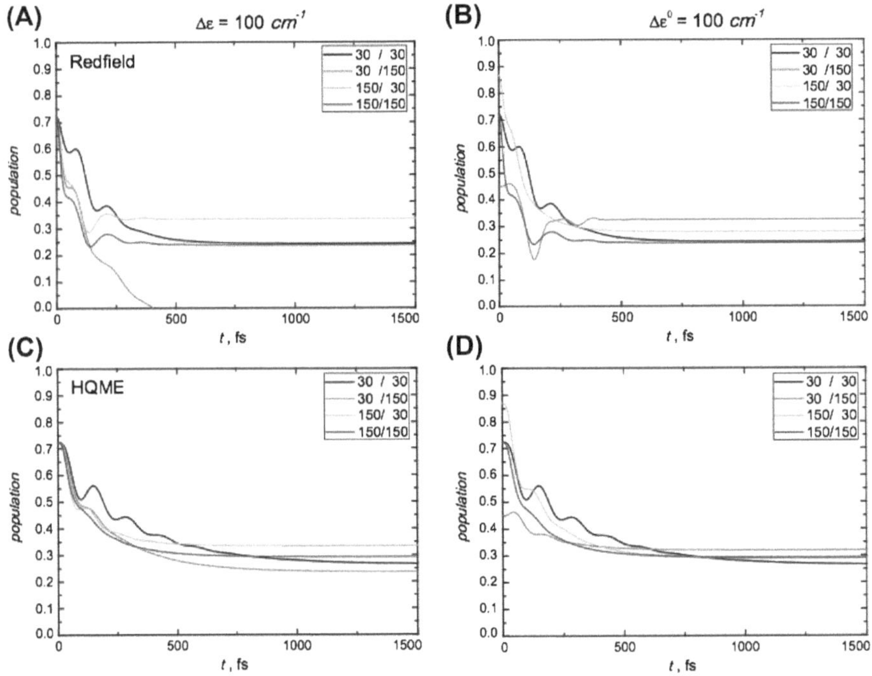

Figure 2 Evolution of the population of the higher excitonic state. The insets show the combination of reorganization energies given in the form λ_a/λ_b (cm^{-1}). The left column corresponds to $\Delta\varepsilon = +100^{-1}$, the right one to $\Delta\varepsilon_0 = +100\ \mathrm{cm}^{-1}$. (A), (B) and (C), (D), correspond to the full Redfield scheme and the HQME equations, respectively.

methods. As we can see, the initial stages of the Redfield and HQME solutions ($\sim$100 fs) look very similar while the rates of the thermalization and the frequencies of the coherent oscillations are different (the coherent oscillations are more dramatic but shorter lived when using the Redfield theory). Moreover, different methods give us different equilibrium population values, and within the HQME approach the latter do not coincide for two pairs of identical λ's.

The case of different reorganization energies (red and green curves) is not trivial (Mancal et al., 2006): the results for both energy gap definitions obtained by all three methods are considerably different in the long time limit. In the right hand column the initial values are scattered simply because fixing $\Delta\varepsilon_0$ with different λ's gives us different $\Delta\varepsilon$ used in the definition of the excitonic basis. Moreover, the red curve in Fig. 2A reveals the well-known problem of the Redfield scheme, namely, that it does not guarantee positivity.

HQME solutions in the excitonic basis, as given in the preceding paragraphs, exhibit nonvanishing steady state coherences. The presence of the steady state coherences can be interpreted as a reflection of renormalization of the system eigenbasis (Schlosshauer, 2007) taking place in the course of time. Contrary to the Redfield scheme, HQME is not subject to any specific predefined basis, even though the equations are originally formulated in the site basis. Therefore, it allows us to identify the so-called preferred basis (or 'global basis' as denoted in Gelzinis et al., 2011) in which the density matrix is diagonal in the long time limit.

The relaxation dynamics described earlier is roughly exponential and can be quantized by thermalization rates. This is especially true when intermolecular couplings are weak. However, the coherent effects may not be neglectable. In this case the Förster rate should approximately describe the dynamics.

The time scale of the excited-state relaxation (or thermalization), τ_{TR}, is obtained exactly by solving the dynamics using the HEOM approach and subsequently performing numerical exponential fitting by means of the least-squares routine. We examine the cases of parameters relevant for the FRET regime, $\lambda \gg J$. Taking $\lambda_a = \lambda_b = 400\ \text{cm}^{-1}$ we consider the cases of $\lambda/J = 20$ ($J = 20\ \text{cm}^{-1}$), which is fully consistent with the FRET regime, and $\lambda/J = 5$($J = 80\ \text{cm}^{-1}$), which reflects the breaking of the validity of FRET. Calculations at 300 K temperature are performed for two different values of environment correlation time $\gamma^{-1} = \gamma_a^{-1} = \gamma_b^{-1} = 10$ and 100 fs. We do not consider the case of a homodimer when the monomeric energy splitting is $\Delta\varepsilon = 0\ \text{cm}^{-1}$.

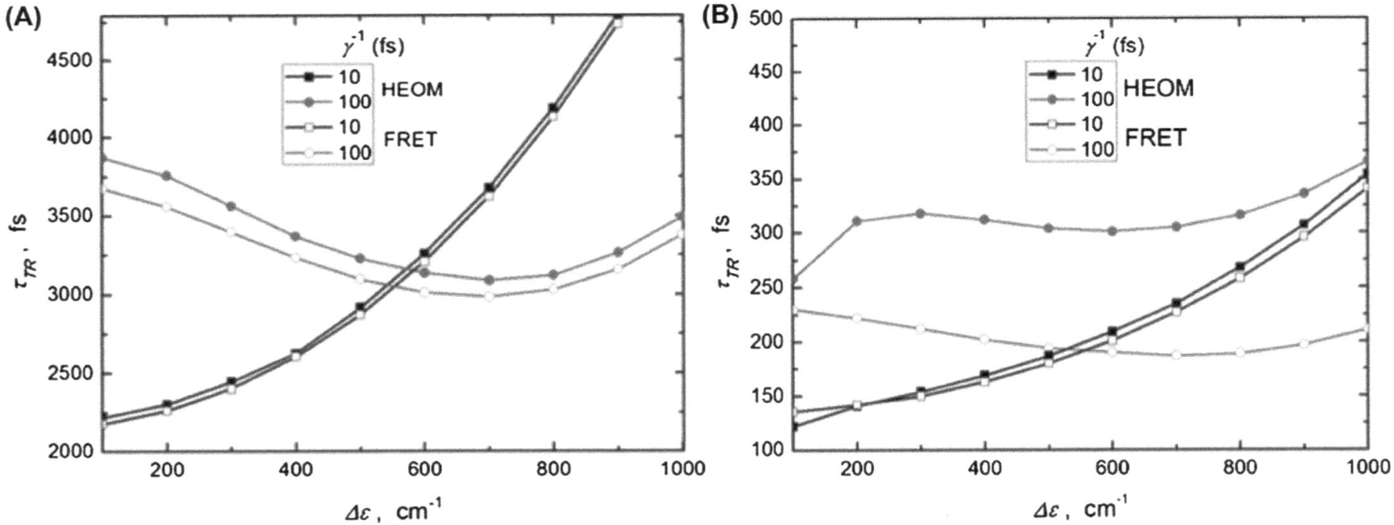

Figure 3 Comparison of electronic relaxation time $\tau_{TR}(\Delta\varepsilon)$ given by HEOM and FRET schemes for different energy gaps. For the sake of consistency, all the rates were extracted from the evolution of the reduced density matrix by numerical fitting. Two cases of resonance coupling correspond to (A) $J = 20\ \text{cm}^{-1}$ and (B) $J = 80\ \text{cm}^{-1}$.

In the case of small resonance coupling and fast correlation decay of the bath (*black squares*, Fig. 3A), the monotonic rise in τ_{TR} is obtained upon the increase of the energy gap. This is an intuitively clear picture of thermalization and corresponds to a Markovian regime of the bath relaxation. In this case, the transfer rates predicted by HEOM and FRET almost match each other (surprisingly, or not, the FRET in the site basis qualitatively describes the Markovian dynamics, governed by the Redfield theory). When we switch to the slow bath (red dots), the dependence changes. Now an optimal system configuration for the excitation transfer around $\Delta\varepsilon \approx 700\ \text{cm}^{-1}$ is well distinguished. This optimal value for transfer rate corresponds to the best matching of energy transfer from the relaxed donor to the unrelaxed acceptor. This is a non-Markovian regime of the bath dynamics, and the HEOM and FRET predictions are qualitatively identical. The FRET approach thus properly describes the parameter regime $\Delta\varepsilon \gg J$ and $\lambda \gg J$. We note that the dependence $\tau_{TR}(\Delta\varepsilon)$ is symmetric around $\Delta\varepsilon = 0$ because the thermalization rate, being the sum of the upward and downward rates is invariant under the change of the sign of $\Delta\varepsilon$. In the case of large resonance coupling, Fig. 3B, the relaxation is substantially faster, as should be expected. Another difference is visible in the HEOM results when energy gaps are $\Delta\varepsilon \leq 300\ \text{cm}^{-1}$. The decrease of $\tau_{TR}(\Delta\varepsilon)$ for smaller energy gaps is the manifestation of the excitonic delocalization effects.

There are two more points to mention which are not shown here explicitly. The first one is that the thermalization rate τ_{TR}^{-1} does not depend on individual reorganization energies but rather on their sum $\lambda_a + \lambda_b$. The second one is that the speedup seen in the non-Markovian regime diminishes upon decreasing $\lambda_a + \lambda_b$ (Balevičius et al., 2012, 2013; Gelzinis et al., 2011).

ACKNOWLEDGEMENT

The research was supported by the Research Council of Lithuania (Grant No. MIP-090/2015).

REFERENCES

Abramavicius, V., & Abramavicius, D. (2014). Excitation transfer pathways in excitonic aggregates revealed by the stochastic Schrödinger equation. *Journal of Chemical Physics, 140*, 065103.

Abramavicius, D., Gulbinas, V., & Valkunas, L. (2003). Acceleration of charge separation by oscillations of the environment polarization. *Chemical Physics Letters, 368*, 480.

Abramavicius, D., & Mukamel, S. (2010). Quantum oscillatory exciton migration in photosynthetic reaction centers. *Journal of Chemical Physics, 133*, 064510.

Abramavicius, D., & Valkunas, L. (2003). Geminate pair recombination in molecular systems with correlated disorder. *Physical Review B, 68*, 245203.

Abramavicius, D., & Valkunas, L. (2016). Role of coherent vibrations in energy transfer and conversion in photosynthetic pigment–protein complexes. *Photosynthesis Research, 127*, 33.

van Amerongen, H., Valkunas, L., & van Grondelle, R. (2000). *Photosynthetic excitons*. Singapore: World Scientific.

Balevičius, V., Jr., Gelzinis, A., Abramavicius, D., Mančal, T., & Valkunas, L. (2012). Excitation dynamics and relaxation in a molecular heterodimer. *Chemical Physics, 404*(0), 94–102.

Balevičius, V., Jr., Gelzinis, A., Abramavicius, D., & Valkunas, L. (2013). Excitation energy transfer and quenching in a heterodimer: applications to the carotenoid–phthalocyanine dyads. *Journal of Physical Chemistry B, 117*(38), 11031–11041.

Bang, O., Christiansen, P. L., If, F., Rasmussen, K. O., & Gaididei, Y. B. (1994). Temperature effects in a nonlinear model of monolayer scheibe aggregates. *Physical Review E, 49*, 4627–4636.

Barzykin, A. V., Frantsuzov, P. A., Seki, K., & Tachiya, M. (2002). Solvent effects in nonadiabatic electron-transfer reactions: theoretical aspects. *Advances in Chemical Physics, 123*, 511.

Basinskaite, E., Butkus, V., Abramavicius, D., & Valkunas, L. (2014). Vibronic models for nonlinear spectroscopy simulations. *Photosynthesis Research, 121*, 95.

Bässler, H. (1993). Charge transport in disordered organic photoconductors a Monte Carlo simulation study. *Physica Status Solidi B, 175*, 15–56.

Beenken, W., Dahlbom, M., Kjellberg, P., & Pullerits, T. (2002). *Journal of Chemical Physics, 117*, 5810.

Blankenship, R. E. (2014). *Molecular mechanisms of photosynthesis* (2nd ed.). Wiley-Blackwell.

Breuer, H.-P., & Petruccione, F. (2002). *The theory of open quantum systems*. New York: Oxford University Press.

Butkus, V., Valkunas, L., & Abramavicius, D. (2011). *Semiconductors and semimetals, v. 85, quantum efficiency in complex systems, part II* (pp. 3–46). San Diego: Academic Press.

Butkus, V., Valkunas, L., & Abramavicius, D. (2012). Molecular vibrations-induced quantum beats in two-dimensional electronic spectroscopy. *Journal of Chemical Physics, 137*, 044513.

Butkus, V., Valkunas, L., & Abramavicius, D. (2014). Vibronic phenomena and exciton–vibrational interference in two-dimensional spectra of molecular aggregates. *Journal of Chemical Physics, 140*, 034306.

Chen, L., Zheng, R., Shi, Q., & Yan, Y. J. (2009). Optical line shapes of molecular aggregates: hierarchical equations of motion method. *Journal of Chemical Physics, 131*(9), 094502.

Cho, M., & Fleming, G. R. (2005). The integrated photon echo and solvation dynamics. II. Peak shifts and 2D photon echo of a coupled chromophore. *Journal of Chemical Physics, 123*, 114506.

Chorosajev, V., Gelzinis, A., Valkunas, L., & Abramavicius, D. (2014). Dynamics of exciton-polaron transition in molecular assemblies: the variational approach. *Journal of Chemical Physics, 140*, 244108.

Dahlbom, M., Beenken, W., Sundström, V., & Pullerits, T. (2002). Collective excitation dynamics and polaron formation in molecular aggregates. *Chemical Physics Letters, 364*, 556–561.

Davydov, A. (1962). *A theory of molecular excitions*. New York: Mc.Graw-Hill.

Diósi, L., Gisin, N., & Strunz, W. T. (1998). Non-markovian quantum state diffusion. *Physical Review A, 58*, 1699–1712.

Diosi, L., & Strunz, W. T. (1997). The non-markovian stochastic schrödinger equation for open systems. *Physics Letters A, 235*, 569–573.

van Eersel, H., Janssen, R. A. J., & Kemerink, M. (2012). Mechanism for efficient photoinduced charge separation at disordered organic heterointerfaces. *Advanced Functional Materials, 22*, 2700–2708.

Förster, T. (1948). Zwischenmolekulare energiewanderung und fluoreszenz. *Annalen der Physik, 6*, 55.

Freiberg, A., Lin, S., Timpmann, K., & Blankenship, R. E. (1997). *Journal of Physical Chemistry B, 101*, 7211.

Frenkel, I., & Frenkel, J. (1936). *Wave mechanics: Elementary theory*. Oxford University Press.

Fuller, F. D., Pan, J., Gelzinis, A., Butkus, V., Senlik, S. S., Wilcox, D. E., … Ogilvie, J. P. (2014). Vibronic coherence in oxygenic photosynthesis. *Nature Chemistry, 6*, 706.

Gazeau, J.-P. (2009). *Coherent states in quantum physics*. KGaA, Weinheim: Wiley-VCH Verlag GmbH & Co.

Gelzinis, A., Abramavicius, D., & Valkunas, L. (2011). Non-Markovian effects in time-resolved fluorescence spectrum of molecular aggregates: tracing polaron formation. *Physical Review B, 84*, 245430.

Gisin, N., & Percival, I. C. (1993). The quantum state diffusion picture of physical processes. *Journal of Physics A: Mathematical and General, 26*(9), 2245.

van Grondelle, R., & Novoderezhkin, V. I. (2006). Energy transfer in photosynthesis: experimental insights and quantitative models. *Physical Chemistry Chemical Physics, 8*, 793–807.

Halls, J. J. M., Walsh, C. A., Greenham, N. C., Marseglia, E. A., Friend, R. H., Moratti, S. C., & Holmes, A. B. (1995). Efficient photodiodes from interpenetrating polymer networks. *Nature, 376*, 498–500.

Heijs, D. J., Dijkstra, A. G., & Knoester, J. (2007). Ultrafast pump-probe spectroscopy of linear molecular aggregates: effects of exciton coherence and thermal dephasing. *Chemical Physics, 341*, 230–239.

Heitele, H. (1993). Dynamic solvent effects on electron-transfer reactions. *Angewandte Chemie International Edition in English, 32*(3), 359–377.

Herrmann, D., Niesar, S., Scharsich, C., Köhler, A., Stutzmann, M., & Riedle, E. (2011). Role of structural order and excess energy on ultrafast free charge generation in hybrid polythiophene/si photovoltaics probed in real time by near-infrared broadband transient absorption. *Journal of the American Chemical Society, 133*, 18220–18233.

Huynh, T., Sun, K.-W., Gelin, M., & Zhao, Y. (2013). Polaron dynamics in two-dimensional photon-echo spectroscopy of molecular rings. *Journal of Chemical Physics, 139*, 104103.

Ishizaki, A., & Fleming, G. R. (2009a). On the adequacy of the redfield equation and related approaches to the study of quantum dynamics in electronic energy transfer. *Journal of Chemical Physics, 130*, 234110.

Ishizaki, A., & Fleming, G. R. (2009b). Unified treatment of quantum coherent and incoherent hopping dynamics in electronic energy transfer: reduced hierarchy equation approach. *Journal of Chemical Physics, 130*, 234111.

Jang, S., Newton, M. D., & Silbey, R. J. (2004). Multichromophoric Förster resonance energy transfer. *Physical Review Letters, 92*, 218301.

Jankowiak, R., Reppert, M., Zazubovich, V., Pieper, J., & Reinot, T. (2011). Site selective and single complex laser-based spectroscopies: a window on excited state electronic structure, excitation energy transfer, and electron phonon coupling of selected photosynthetic complexes. *Chemical Reviews, 111*, 4546.

van Kampen, N. G. (2007). *Stochastic processes in physics and chemistry* (3rd ed.). North Holland.

Keivanidis, P. E., Clarke, T. M., Lilliu, S., Agostinelli, T., Macdonald, J. E., Durrant, J. R., … Nelson, J. (2010). Dependence of charge separation efficiency on film microstructure in poly(3-hexylthiophene-2,5-diyl):[6,6]-phenyl-c 61 butyric acid methyl ester blend films. *Journal of Physical Chemistry Letters, 1*, 734–738.

Kell, A., Feng, X., Reppert, M., & Jankowiak, R. J. (2013). On the shape of the phonon spectral density in photosynthetic complexes. *Journal of Physical Chemistry B, 117*, 7317–7323.

Kjellberg, P., Brüggemann, B., & onu Pullerits, T. (2006). Two-dimensional electronic spectroscopy of an excitonically coupled dimer. *Physical Review B, 74*, 024303.

Lee, J., Vandewal, K., Yost, S. R., Bahlke, M. E., Goris, L., Baldo, M. A., ... Voorhis, T. V. (2010). Charge transfer state versus hot exciton dissociation in polymer-fullerene blended solar cells. *Journal of the American Chemical Society, 132*, 11878–11880.

Mančal, T., Balevičius, V., & Valkunas, L. (2011). Decoherence in weakly coupled excitonic complexes. *Journal of Physical Chemistry A, 115*, 3845–3858.

Mancal, T., Valkunas, L., & Fleming, G. R. (2006). Theory of exciton-charge transfer state coupled systems. *Chemical Physics Letters, 432*, 301–305.

Marcus, R. A. (1963). On the theory of oxidation-reduction reactions involving electron transfer. v. comparison and properties of electrochemical and chemical rate constants. *Journal of Physical Chemistry, 67*, 853–857.

Marcus, R., & Sutin, N. (1985). Electron transfers in chemistry and biology. *Biochimica et Biophysica Acta (BBA) – Reviews on Bioenergetics, 811*(3), 265–322.

Marsh, R. A., Hodgkiss, J. M., & Friend, R. H. (2010). Direct measurement of electric field-assisted charge separation in polymer: fullerene photovoltaic diodes. *Advanced Materials, 22*, 3672–3676.

May, V., & Kühn, O. (2011). *Charge and energy transfer dynamics in molecular systems*. Weinheim: Wiley-VCH.

Mukamel, S. (1995). *Principles of nonlinear optical spectroscopy*. New York: Oxford University Press.

O'Sullivan, M. C., Sprafke, J. K., Kondratuk, D. V., Rinfray, C., Claridge, T. D. W., Saywell, A., ... Anderson, H. L. (2011). Vernier templating and synthesis of a 12-porphyrin nano-ring. *Nature, 469*(7328), 72–75.

Panitchayangkoon, G., Voronine, D., Abramavicius, D., Caram, J., Lewis, N., Mukamel, S., & Engel, G. (2011). Direct evidence of quantum transport in photosynthetic light-harvesting complexes. *Proceedings of the National Academy of Sciences of the United States of America, 108*, 20908–20912.

Piilo, J., Härkönen, K., Maniscalco, S., & Suominen, K.-A. (2009). Open system dynamics with non-markovian quantum jumps. *Physical Review A, 79*, 062112.

Piilo, J., Maniscalco, S., Härkönen, K., & Suominen, K.-A. (2008). Non-markovian quantum jumps. *Physical Review Letters, 100*, 180402.

Pisliakov, A. V., Mancal, T., & Fleming, G. R. (2006). Two-dimensional optical three-pulse photon echo spectroscopy. ii. signatures of coherent electronic motion and exciton population transfer in dimer two-dimensional spectra. *Journal of Chemical Physics, 124*, 234505.

Polyutov, S., Kühn, O., & Pullerits, T. (2012). Exciton-vibrational coupling in molecular aggregates: electronic versus vibronic dimer. *Chemical Physics, 394*, 21–28.

Pope, M., & Swenberg, C. E. (1999). *Electron processes in organic crystals*. New York/Oxford: Oxford University Press.

Rancova, O., Jankowiak, R., & Abramavicius, D. (2015). Probing environment fluctuations by two-dimensional electronic spectroscopy of molecular systems at temperatures below 5 K. *Journal of Chemical Physics, 142*, 212428.

Redfield, A. G. (1957). On the theory of relaxation processes. *IBM Journal of Research and Development, 1*, 19–31.

Renger, T. (2004). Theory of optical spectra involving charge transfer states: dynamic localization predicts a temperature dependent optical band shift. *Physical Review Letters, 93*, 188101.

Ruban, A. V. (2012). *The photosynthetic membrane: Molecular mechanisms and biophysics of light harvesting*. Wiley.
Schlau-Cohen, G. S., Ishizaki, A., & Fleming, G. R. (2011). Two-dimensional electronic spectroscopy and photosynthesis: fundamentals and applications to photosynthetic light-harvesting. *Chemical Physics, 386*, 1–22.
Schlosshauer, M. (2007). *Decoherence and the quantum-to-classical transition. The frontiers collection*. Berlin: Springer Verlag.
Scot, A. (1992). *Physics Reports, 217*, 1.
Song, A., & Williams, R. (1993). Self-trapped excitons. *Springer Series in Solid-State Sciences, 105*.
Sumi, H., & Marcus, R. A. (1986). Dynamical effects in electron transfer reactions. *Journal of Chemical Physics, 84*(9), 4894–4914.
Sun, J., Luo, B., & Zhao, Y. (2010). Dynamics of a one-dimensional Holstein polaron with the Davydov ansätze. *Physical Review B, 82*, 014305.
Tanimura, Y. (2006). Stochastic liouville, langevin, fokker-planck, and master equation approaches to quantum dissipative systems. *Journal of the Physical Society of Japan, 75*, 082001.
Timpmann, K., Rätsep, M., Hunter, C. N., & Freiberg, A. (2004). Emitting excitonic polaron states in core lh1 and peripheral lh2 bacterial light-harvesting complexes. *Journal of Physical Chemistry B, 108*(29), 10581–10588.
Valkunas, L., Abramavicius, D., & Mancal, T. (2013). *Molecular excitation dynamics and relaxation*. Weinheim: Wiley-VCH.
Vithanage, D. A., Devižis, A., Abramavičius, V., Infahsaeng, Y., Abramavicius, D., MacKenzie, R., … Gulbinas, V. (2013). Visualizing charge separation in bulk heterojunction organic solar cells. *Nature Communications, 4*, 2334.
Warshel, A., & Parson, W. W. (2001). Dynamics of biochemical and biophysical reactions: insight from computer simulations. *Quarterly Reviews of Biophysics, 34*, 563.
Weiss, U. (2008). *Quantum dissipative systems*. World Scientific.
Xu, R. X., Tian, B. L., Xu, J., Shi, Q., & Yan, Y. J. (2009). Hierarchical quantum master equation with semiclassical drude dissipation. *Journal of Chemical Physics, 131*, 214111.
Ye, J., Sun, K., Zhao, Y., Yu, Y., Kong Lee, C., & Cao, J. (2012). Excitonic energy transfer in light-harvesting complexes in purple bacteria. *Journal of Chemical Physics, 136*(24), 245104.
Yu, T., Diósi, L., Gisin, N., & Strunz, W. T. (1999). Non-markovian quantum-state diffusion: perturbation approach. *Physical Review A, 60*, 91–103.
Yu, G., Gao, J., Hummelen, J. C., Wudl, F., & Heeger, A. J. (1995). Polymer photovoltaic cells: enhanced efficiencies via a network of internal donor-acceptor heterojunctions. *Science, 270*, 1789–1791.
Zhang, W.-M., Feng, D. H., & Gilmore, R. (1990). Coherent states: theory and some applications. *Reviews of Modern Physics, 62*, 867–927.

CHAPTER SIX

Resolving Energy and Electron Transfer Processes in Dyads With the Help of Global and Target Analysis[a]

I.H.M. van Stokkum*,[1], J. Ravensbergen*, J.J. Snellenburg*, R. van Grondelle*, S. Pillai§, T.A. Moore§, D. Gust§, A.L. Moore§, J.T.M. Kennis*

*Vrije Universiteit, Amsterdam, The Netherlands
§Arizona State University, Tempe, AZ, United States
[1]Corresponding author: E-mail: i.h.m.van.stokkum@vu.nl

Contents

Abstract

In any photosynthetic/photocatalytic device, multiple steps are required between the arrival of a solar photon and the formation of a stable product. Here we explain and demonstrate the target analysis methodology to develop minimal models, identify the steps and estimate the parameters that characterize energy converting devices. With this modelling tool the molecular mechanisms of the loss processes can be identified and quantified. This can then inspire photosynthetic device optimization by

[a] This chapter extends the data analysis and modelling of Pillai et al. (2013).

Advances in Botanical Research, Volume 79
ISSN 0065-2296
http://dx.doi.org/10.1016/bs.abr.2016.04.003

precisely targeting those sites involved in the most significant losses. Two case studies of recently published measurements (Pillai et al., 2013) on a carotenoporphyrin dyad and a carotenofullerene dyad are modelled in depth. After carotenoid excitation, no excited state energy transfer (EET) to porphyrin was found, but EET from carotenoid hot S_1 to the fullerene moiety occurred with a rate of 1.6/ps. The total radical pair yields of these dyads were found to be, respectively, 46% and 79%. Out of these 79%, 31% were due to electron transfer from the fullerene excited state. The triplet yields were 3.8% and 4.6%. The remainder of the excitations decay to the ground state from the carotenoid hot S_1 and S_1 states.

List of Abbreviations

Car	Carotenoid
DADS	Decay-associated difference spectra
EADS	Evolution-associated difference spectra
EET	Excited state energy transfer
ESA	Excited state absorption
ET	Electron transfer
Ful	Fullerene
FWHM	Full width at half maximum
IC	Internal conversion
ICT	Intramolecular charge-transfer
IRF	Instrument response function
NIR	Near infrared
rms	Root mean square
RP	Radical pair
SADS	Species-associated difference spectra
SE	Stimulated emission

1. INTRODUCTION

In any photosynthetic/photocatalytic device, multiple steps are required between the arrival of a solar photon and the formation of a stable product: light absorption, exciton migration, charge separation, charge transport and the catalytic events. Each step is associated with unavoidable losses and as a consequence the overall process may have a low yield. To optimize the functioning of a prototype photocatalytic/photosynthetic device, these loss processes must be identified rigorously. A model-based description of natural, hybrid or artificial photosynthetic devices should identify and quantify the states and the losses and thereby facilitate a quantitative comparison of the different devices. To assess the efficiency of a photosynthetic/photocatalytic device, most often a variety of time-resolved spectroscopic (Vis-IR, ESR etc.) measurements are employed to monitor the dynamics of the light absorption, energy transfer, charge separation, charge transport and catalysis. Time-resolved emission spectroscopy monitors the fate of the excited states, whereas time-resolved difference absorption spectroscopy can also probe nonradiative states, in particular charge separates states. Reviews of ultrafast measurements (fs, ps, ns time scales) are (Berera, van Grondelle, & Kennis, 2009; Fleming, 1986; Holzwarth, 1995; Polívka & Sundstrom, 2004).

In every step, loss processes can occur, leading to a decrease in efficiency or unwanted by-products. To study the functioning of a device, a quantitative model that describes all time-resolved spectroscopic measurements is mandatory. Typically, the models consist of differential equations containing unknown kinetic parameters (k_{yield} and k_{loss}) and unknown spectral parameters that characterize molecular states (1, 2,..., n, product) (see Fig. 1). The target analysis methodology (Holzwarth, 1996; van Stokkum, Larsen, & van Grondelle, 2004; van Stokkum, van Oort, van Mourik, Gobets, & van

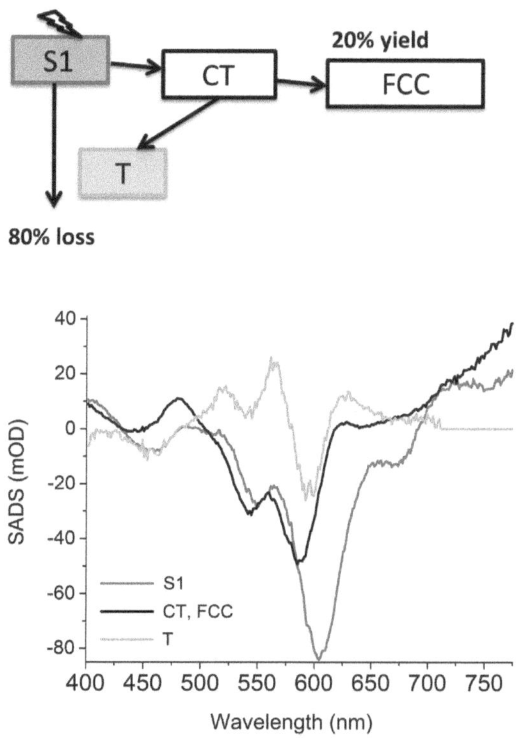

Figure 1 Kinetic scheme of an organic solar cell with species-associated difference spectra (SADS) of four states (Williams, Vân Anh, & van Stokkum, 2013). Excited state S_1 (dark grey) produces a charge transfer state (CT, black) which yields free charge carriers (FCC, black). The losses through charge recombination result in triplets (T, light grey). A target analysis identifies the states involved and quantifies the losses.

Amerongen, 2008) is necessary to develop minimal models, identify the states by means of their species-associated difference spectra (SADS) and estimate the parameters that characterize energy converting devices. With this modelling tool the molecular mechanisms of the loss processes can be identified and quantified. This can then inspire photosynthetic device optimization by precisely targeting those sites involved in the most significant losses.

In this chapter we will reanalyse and model measurements of molecular donor—acceptor systems (dyads) (Pillai et al., 2013). The emphasis will be on the data analysis methodology. For the details on design, synthesis, experiment and photophysics, the reader is referred to the original publication.

Photophysical investigations of molecular donor—acceptor systems have elucidated many details of natural photosynthesis and revealed design principles for artificial photosynthetic systems (Berera, Herrero, et al., 2006; Berera, Moore, et al., 2006; Fukuzumi, 2006; Guldi, 2002; Kloz et al., 2011, 2012; Kodis et al., 2004; Williams, Zwier, & Verhoeven, 1995). To obtain insights into the factors that govern the partition between excited state energy transfer and electron transfer (ET) processes among carotenoids and tetrapyrroles or fullerenes, Pillai et al. (2013) have designed artificial photosynthetic dyads that are thermodynamically poised to favour ET over EET processes. Two biomimetic artificial photosynthetic constructs were synthesized and investigated: a carotenoporphyrin dyad (dyad **1**) and a carotenofullerene dyad (dyad **2**), as depicted in Fig. 2.

Dyad **1** consists of a carotenoid covalently attached to a porphyrin moiety, substituted with electron withdrawing CN groups to enhance its electron accepting properties. The carotenoid has nine double bonds in the polyene chain, an additional double bond in the β-ionone ring and an elongation of the conjugated system into the linker moiety by a phenyl group. The effective conjugation length of this carotenoid is estimated at 11 double bonds (considering the phenyl group as 1.5 conjugation length and the cyclohexene-like β-ionone ring as 0.5 conjugation length, because it is rotated out of the plane of the polyene). The bulky substituted phenyl groups are oriented at steep angles to the porphyrin plane and are added to prevent the dyad from stacking. The two pigments are linked through two phenyl rings and a peptide bond.

Dyad **2** has a carotenoid with eight double bonds in the polyene chain and one in the adjacent β-ionone ring, which gives an effective conjugation length of 8.5 double bonds. It is coupled to a C_{60} by a positively charged methylated pyrrolidine linker. Fullerenes are not found in biological systems but are promising components for artificial photosynthesis (Gust, Moore, &

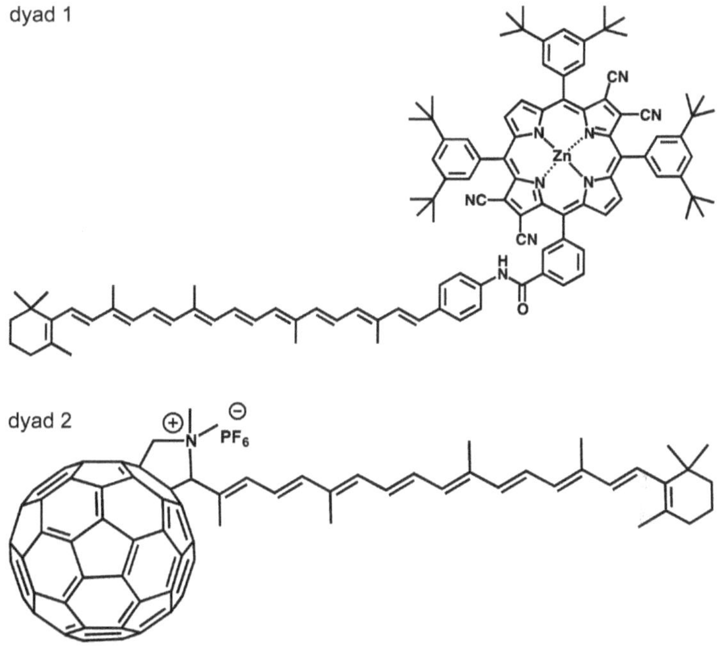

Figure 2 Structure of the dyads: carotenoporphyrin (dyad 1) and carotenofullerene (dyad 2).

Moore, 2000; Wróbel & Graja, 2011). They have a small reorganization energy in ET reactions, which can help place the ET reaction at the peak of the Marcus curve, facilitating fast photoinduced ET and slow recombination (Hiroshi et al., 1996; Marcus & Sutin, 1985).

Both dyads were studied by transient difference absorption spectroscopy, measuring the time-resolved difference spectrum after excitation of the carotenoid moiety. Global analysis was used to describe the data with a minimal set of lifetimes and associated spectra. Subsequently, in a target analysis a detailed kinetic model including EET and ET pathways was tested, and the corresponding rates and species-associated spectra were estimated. For dyad **1**, Pillai et al. (2013) found that ET from the carotenoid S_2 state to the porphyrin completely outcompetes any EET process. In dyad **2**, ET from carotenoid to the fullerene is the dominant process, but an EET process from the optically forbidden carotenoid S_1 state to the lowest fullerene singlet-excited state is observed as well. Here we test their interpretation, propose detailed kinetic schemes and quantify all photophysical processes in terms of the spectra of the states involved and the rate constants describing their interconversion.

2. METHODOLOGY

2.1 Model-Based Data Analysis

In time-resolved spectroscopic experiments, a very large amount of data is collected which can be analysed by global and target analysis techniques. A time-resolved experiment $\Delta A(t, \lambda)$ typically consists of time-gated spectra at several 100 delays, each spectrum consisting of several 100 wavelengths, resulting in a collection of tens of thousands of data points. The aim of the data analysis is to obtain a model-based description of the full data set in terms of a model containing a small number of precisely estimated parameters, of which the rate constants and spectra are the most relevant. A description of the basic ingredient of kinetic models, the exponential decay, will be given first, followed by a description of how to use these ingredients for global and target analysis of the full data. Our main assumption here is that the time and wavelength properties of the system of interest are separable, which means that spectra of species or states are constant. For details on parameter estimation techniques, the reader is also referred to I. H. M. van Stokkum et al. (2004). Software issues are discussed in I. H. M. van Stokkum and Bal (2006).

2.2 Modelling an Exponential Decay

Here an expression is derived for describing an exponentially decaying component. The instrument response function (IRF) that accounts for the convolution of excitation, and probe pulse is usually modelled with a Gaussian parameterized using μ and Δ for, respectively, location and full width at half maximum (FWHM):

$$\mathrm{IRF}(t) = \frac{1}{\tilde{\Delta}\sqrt{2\pi}} \exp\left(-\log(2)(2(t-\mu)/\Delta)^2 \right)$$

where $\tilde{\Delta} = \Delta/(2\sqrt{2\log(2)})$. Typically the FWHM is ≈ 100 fs in the visible and ≈ 200 fs in the near-IR. The convolution (indicated by an $\star$) of this IRF with an exponential decay (with rate k) yields an analytical expression which facilitates the estimation of the decay rate k and the IRF parameters μ and Δ:

$$\begin{aligned} c^I(t, k, \mu, \Delta) &= \exp(-kt) * \mathrm{IRF}(t) \\ &= \frac{1}{2}\exp(-kt)\exp\left(k\left(\mu + \frac{k\tilde{\Delta}^2}{2}\right)\right) \\ &\quad \times \left\{1 + \mathrm{erf}\left(\frac{t - (\mu + k\tilde{\Delta}^2)}{\sqrt{2}\tilde{\Delta}}\right)\right\} \end{aligned}$$

By the nature of the white light generation process, the white light is 'chirped' on generation, ie, the 'blue' wavelengths are generated later in time than the 'red' wavelengths. Hence, the white light continuum has an intrinsic group velocity dispersion. When travelling through optically dense materials such as lenses and cuvettes, the group velocity dispersion in the white light readily increases to picoseconds. The group velocity dispersion can be modelled with a polynomial function that describes the wavelength dependence of the IRF location μ:

$$\mu(\lambda) = \mu_{\lambda_c} + \sum_{j=1}^{j_{\max}} a_j(\lambda - \lambda_c)^j$$

Typically, with visible difference absorption measurements over a large wavelength range, the order of this polynomial ($j_{\max}$) is three. The reference wavelength λ_c is conveniently at the centre of the spectrograph. In addition, also the IRF width can be wavelength dependent, and an analogous polynomial description can be employed.

2.3 Global and Target Analysis

The basis of global analysis is the superposition principle, which states that the measured data $\Delta A(t, \lambda)$ result from a superposition of the spectral properties $\varepsilon_l(\lambda)$ of the components present in the system of interest weighted by their concentration $c_l(t)$.

$$\Delta A(t, \lambda) = \sum_{l=1}^{n_{\text{comp}}} c_l(t)\varepsilon_l(\lambda)$$

The $c_l(t)$ of all n_{comp} components are described by a compartmental model, that consists of coupled first-order differential equations, with as solution sums of exponential decays. We will consider three types of compartmental models: (1) a model with components decaying monoexponentially in parallel, which yields decay-associated difference spectra (DADS), (2) a sequential model with increasing lifetimes, also called an unbranched unidirectional model (Nagle, Parodi, & Lozier, 1982), resulting in evolution-associated difference spectra (EADS) and (3) a full compartmental scheme which may include possible branchings and equilibria, yielding SADS. The latter is most often referred to as target analysis, where the target is the proposed kinetic scheme, including possible spectral assumptions. When analysing fluorescence data (which are nonnegative) the word *Difference* is not applicable, and the nomenclature is DAS, EAS and SAS.

With parallelly decaying components, the model reads

$$\Delta A(t,\lambda) = \sum_{l=1}^{n_{\text{comp}}} c^I(k_l)\text{DADS}_l(\lambda)$$

The DADS thus represent the estimated amplitudes of the above-defined exponential decays $c^I(k_l)$. When the system consists of parallelly decaying components, the DADS are true species difference spectra. In all other cases, they are interpreted as a weighted sum (with both positive and negative contributions) of the true SADS.

In addition to exponentially decaying components, scattering or a coherent artefact can be present. This can often be modelled as an extra component with as time profile the IRF shape. Such spectra are typical in nonlinear optics and can also contain Raman scattering from the solvent (Berera et al., 2009).

Following a successfully converged fit, the matrix of residuals is then analysed with the help of a singular value decomposition (SVD). Formally the residual matrix can be decomposed as

$$\text{res}(t,\lambda) = \sum_{l=1}^{m} u_l^{\text{res}}(t)s_l w_l^{\text{res}}(\lambda)$$

where u_l and w_l are the left and right singular vectors, s_l the sorted singular values and m is the minimum of the number of rows and columns of the matrix. The singular vectors are orthogonal and provide an optimal least squares approximation of the matrix. The SVD is useful to diagnose shortcomings of the model used or systematic errors in the data.

A sequential model consists of components decaying sequentially $1 \rightarrow 2 \rightarrow \ldots \rightarrow n_{\text{comp}}$ where the decay rates k_l are decreasing, and the lifetimes $1/k_l$ are increasing. It is a convenient way to describe the spectral evolution, and each component possesses an EADS. The model for the data now reads

$$\Delta A(t,\lambda) = \sum_{l=1}^{n_{\text{comp}}} c_l^{II}\text{EADS}_l(\lambda)$$

where each concentration is a linear combination of the exponential decays, $c_l^{II} = \sum_{j=1}^{l} b_{jl}c^I(k_l)$, and the amplitudes (Nagle et al., 1982) b_{jl} are given by $b_{11} = 1$ and for $j \le l$:

$$b_{jl} = \prod_{m=1}^{l-1} k_m \Big/ \prod_{n=1,n\neq j}^{l} (k_n - k_j)$$

When the system truly consists of sequentially decaying components $1 \rightarrow 2 \rightarrow \ldots \rightarrow n_{\text{comp}}$, the EADS are true SADS (see later discussion). In all other cases, they are interpreted as a weighted sum (with only positive contributions) of SADS.

In a more general target analysis a compartmental scheme is used to describe the concentrations of the compartments (states). Transitions to and from compartments are described by microscopic rate constants. The model that is fitted to the data now reads

$$\Delta A(\lambda, t) = \sum_{i=1}^{n_{\text{comp}}} c_i(t)\text{SADS}_i(\lambda)$$

where $c_i(t)$ now corresponds to the concentration of the i-th compartment. $\text{SADS}_i(\lambda)$ is the i-th SADS. The concentrations of all compartments are collated in a vector. When a carotenoid is excited, which ultimately results in a product, such a vector reads: $c(t) = [c_1(t)\ c_2(t) \ldots\ \ c_{n_{\text{comp}}}(t)]^T = [\text{Car}^*(t)\ \ \ldots\ \ \text{Intermediate}(t)\ \ \ldots\ \ \text{Product}(t)]^T$ which obeys the differential equation

$$\frac{d}{dt}c(t) = Kc(t) + j(t)$$

where the transfer matrix K contains off-diagonal elements k_{pq}, representing the microscopic rate constant from compartment q to compartment p. The diagonal elements contain the total decay rates of each compartment. The input to the compartments is $j(t) = \text{IRF}(t)[1\quad 0\quad 0]^T$. In most cases spectral constraints are needed to limit the number of parameters and render the target model identifiable.

A software package called TIMP (Mullen & van Stokkum, 2007) which implements these models was developed for the statistical computing language R (R Development Core Team, 2010) and a corresponding sophisticated graphical user interface called Glotaran has been released (Snellenburg, Laptenok, Seger, Mullen, & van Stokkum, 2012).

3. APPLICATIONS TO SYNTHETIC SYSTEMS

The absorption spectra of dyads **1** and **2** in the nonpolar solvent toluene (dielectric constant 2.4) are shown in Fig. 3, in black and grey respectively. With dyad **1**, the double band around 450 nm corresponds to the porphyrin Soret band, the bands around 660 and 680 nm to the

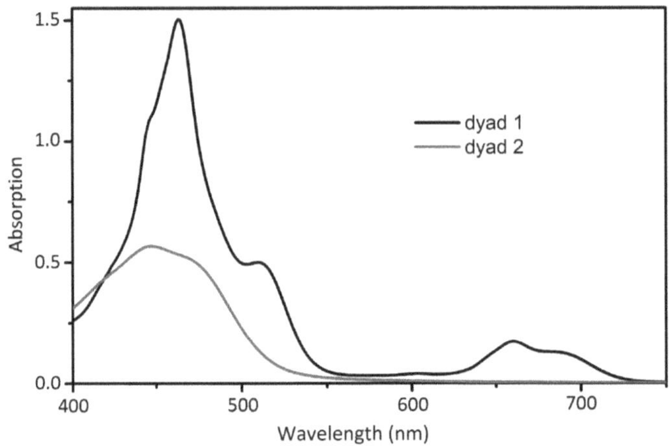

Figure 3 Absorption spectra. Black: dyad 1 in toluene. Grey: dyad 2 in toluene. All spectra were measured in a 1 mm path length quartz cuvette.

porphyrin Q-band absorption. The carotenoid absorption bands overlap with the porphyrin Soret band, except for the lowest vibronic S_2 transition, which is found at 515 nm.

With dyad **2**, the carotenoid absorption is shifted to the blue with respect to dyad **1**, as expected from the shorter conjugation length of the molecule, leading to a larger energy gap between the ground and excited states (Polívka & Sundstrom, 2004). Fullerene has an absorption band around 300 nm (Guldi & Prato, 2000) (not shown) and an absorption tail with very small amplitude that extends throughout the visible range.

3.1 Dyad 1

Dyad **1** (Fig. 2) has been excited in toluene solution in its carotenoid moiety at $\approx$515 nm. The data from transient absorption measurements have been globally analysed using a simultaneous analysis for measurements in the visible and near-IR. Fig. 4 contains the EADS for dyad **1** in toluene. At least six components are needed for a sufficient fit of the data. The estimation of the fastest process crucially depends upon an adequate description of the wavelength dependence of the IRF. Both the location and width of the IRF are wavelength dependent. With six components (lifetimes 78 fs, 1.1 ps, 4.5 ps, 243 ps, 810 ps, long lived) the overall rms error of the fit is 0.555, but a trend is visible in the first left singular vector of the residual matrix of the NIR data. Addition of a seventh component (lifetimes 30 fs

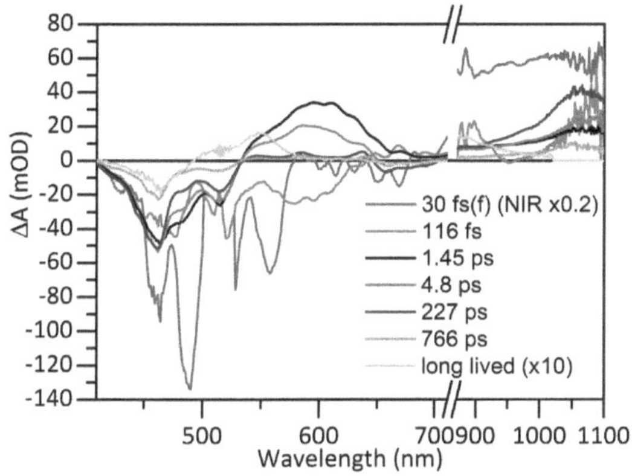

Figure 4 EADS of dyad 1 in toluene after 515 nm excitation.

(fixed), 116 fs, 1.45 ps, 4.8 ps, 227 ps, 766 ps, long lived) results in a significant decrease of the overall rms error of the fit to 0.524, and the above-mentioned trend is no longer visible. Here we present this seven component fit. In the visible, due to dispersion, the IRF location increased by 0.83 ps between 410 and 710 nm. In the near-IR, the IRF location increased by 0.26 ps between 830 and 1120 nm. Concomitantly, the FWHM of the IRF decreased from to 0.19 to 0.12 ps in the visible and increased from 0.31 to 0.38 ps in the near-IR. A broadening of the IRF towards 410 nm results from the higher energy of the probe light. The first lifetime of 30 fs was fixed, to summarize the relaxation of the S_2 state. Next, a second lifetime of 116 fs was estimated, that could be attributed to the decay of the relaxed carotenoid S_2 state. We will first discuss the EADS, since they correspond to a spectral evolution that can be intuitively understood, and later discuss the DADS, which are more difficult to interpret. The first EADS (grey) shows negative ΔA bands in the visible and a positive band in the near-IR region (Note that the latter is much larger that the later signals in the near-IR and was multiplied by 0.2 for comparison.). This is the typical absorbance difference spectrum due to population of the carotenoid S_2 state and is characterized by ground state bleach and stimulated emission (negative ΔA signals) in the visible and excited state absorption (positive ΔA signal) in the near-IR(Zhang, Inaba, Watanabe, & Koyama, 2000a, 2000b). The minimum at 490 nm can be attributed to Raman scatter of the toluene solvent, which is maximal at 1000/cm (Hitchcock & Lapiosa, 1975). This energy corresponds to the difference between the 515 nm excitation wavelength and 490 nm ($10^7/490 - 10^7/515 \approx 1000/\text{cm}$). Note that the SE from S_2 to S_0 is largest at 558 nm. This location red shifts to 579 nm in the second EADS (green), which largely represents the relaxed carotenoid S_2 state. Moreover, in the NIR the EADS can be interpreted as an SE from S_2 to S_x superimposed upon a flat ESA. Thus this EADS is minimal near 950 nm, due to the opposite contributions of ESA and SE. From the difference between these minima, we estimate the energy of the S_x state as $10^7/579 - 10^7/950 \approx 6750$/cm. This is reminiscent of the peridinin carotenoid where (Zigmantas, Polívka, Hiller, Yartsev, & Sundström, 2001) found SE from an intramolecular charge-transfer state, with SE around 1000 nm. These ultrafast carotenoid S_2 dynamics on the 100 fs time scale deserve further investigation. In the remainder we concentrate on the carotenoid porphyrin interactions. The second EADS (green) evolves into the third EADS (black) with a time constant of 116 fs.

Both the third (black, 1.45 ps) and fourth (red, 4.8 ps) EADS show a positive band between 530 and 700 nm, originating from carotenoid S_1 excited state absorption (Koyama, Kakitani, Miki, Christiana, & Nagae, 2010; Polívka & Sundstrom, 2004). The excited state absorption of the fourth EADS is blue-shifted with respect to the third, which corresponds to a transition from vibrationally hot to relaxed S_1 (Koyama et al., 2010; Polívka & Sundstrom, 2004). Both the 1.45 ps EADS and the 4.8 ps EADS show a negative signal below 530 nm, which can be interpreted as a mix of carotenoid and porphyrin Soret bleaching. In the near-IR, these EADS show a band at 1060 nm that can be assigned to the carotenoid cation radical (Jeevarajan, Wei, Jeevarajan, & Kispert, 2011; Kodis et al., 2004). This, together with the presence of porphyrin bleaching, indicates that charge separation has taken place. Hence we can attribute these two spectra to a mix of carotenoid S_1 (or hot S_1) and the charge-separated $Car^{\bullet +}$ $Por^{\bullet -}$ state.

The fifth (blue, 227 ps) EADS shows bleaching signatures around 650 nm originating from porphyrin and around 450 and 500 nm originating from both porphyrin and carotenoid. Furthermore, the spectrum shows an increased absorption of the carotenoid cation radical in the near-IR. The cation absorption band has increased with respect to the fourth spectrum (red 4.8 ps), while the carotenoid S_1 absorption band has decreased. This observation shows that charge separation has taken place from the S_1 state in 4.8 ps. The 766 ps EADS (magenta line) has a spectral shape that is virtually identical to that of the 227 ps EADS, but with significantly reduced amplitude, and hence the 227 and 766 ps time constants are assigned to radical pair recombination. The need for two recombination time constants is probably related to heterogeneity in the sample or the typical

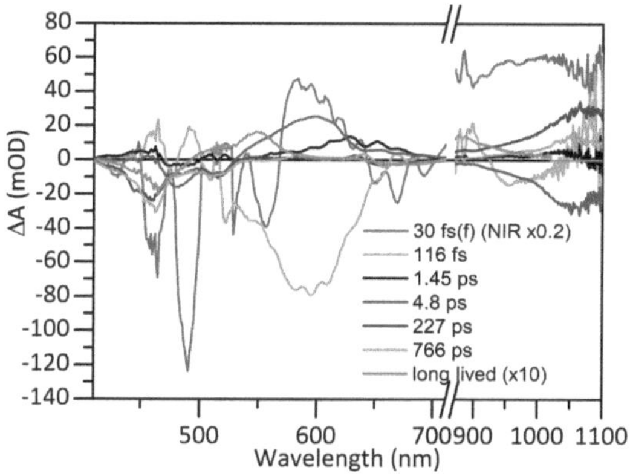

Figure 5 DADS of dyad 1 in toluene.

multiexponential decay reflecting disorder. Structural heterogeneity has been demonstrated previously in axially linked caroteno-phthalocyanine triads by NMR spectroscopy (Palacios et al., 2006). Alternatively, there is a 227 ps evolution towards a state with less carotenoid cation radical absorption (see later discussion).

The last EADS (cyan, long lived) is nondecaying on the time scale of this experiment. The spectrum has been expanded 10× for clarity. It shows an excited state absorption band at 550 nm. This band is characteristic of the long-lived carotenoid triplet state (Berera, Herrero, et al., 2006; Berera, Moore, et al.,2006; Koyama et al., 2010; Polívka & Sundstrom, 2004). Usually, this band is only accompanied by carotenoid bleaching. Here, we also see porphyrin bleaching at both the Q and Soret wavelengths. The presence of porphyrin features in the final state is surprising, as the charge-separated state has recombined in the previous processes, and porphyrin triplet states are, in general, energetically higher than those of carotenoids. As observed previously in several natural and artificial light-harvesting systems, this observation suggests that the triplet state is delocalized on the carotenoid and the porphyrin, which leads to a ground state bleach signal of both moieties (Alexandre et al., 2007; Angerhofer, Bornhäuser, Gall, & Cogdell, 1995; Bonetti et al., 2010; Gall et al., 2011).

To complement this EADS discussion we now describe some of the DADS features, cf. Fig. 5. Recall that the quality of the fit is identical with a sequential or parallel model; however, the DADS provide a differential view on the spectral evolution, stressing the changes.

The first DADS (grey) shows a decrease of the 550 nm SE and an increase of the 580 nm SE, that correspond to the relaxation of the S_2 state.

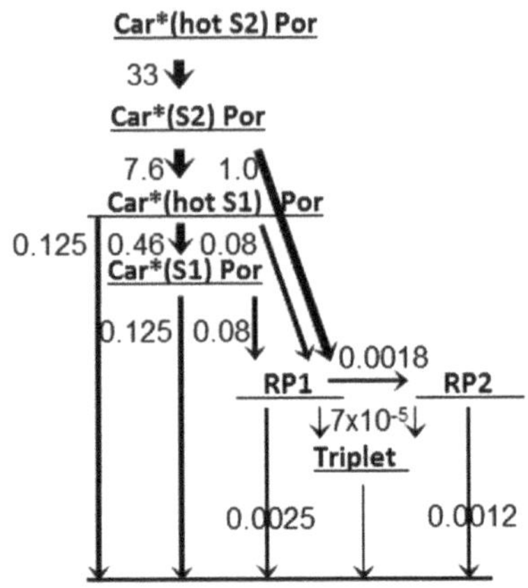

Figure 6 Kinetic scheme for dyad 1 with estimated rate constants in 1/ps. *Car*, Carotenoid; *RP*, radical pair.

The second DADS (green) shows a large negative band from 500 to 650 nm. This can be attributed to the decay of the carotenoid S_2 stimulated emission and the rise of the carotenoid S_1 excited state absorption. In addition a small positive band near 450 nm indicates rise of the porphyrin bleaching. In the NIR the negative band around 950 nm reflects SE from relaxed S_2 to S_x superimposed upon a flat ESA. The third DADS (1.45 ps) shows positive bands from 600 to 700 nm and from 1000 to 1070 nm, which can be attributed to the relaxation of the vibrationally hot S_1. The negative band at 1060 nm of the 4.8 ps DADS (red) indicates the main rise of the carotenoid cation radical. The differences between the 227 and 766 ps DADS are subtle in the visible. The porphyrin bleaching near 450 nm is stronger in the 227 ps DADS, and porphyrin signals near 660 nm are strongest in the 766 ps DADS. In the near-IR the main decay of the 1060 nm carotenoid cation radical band is with 227 ps. Strikingly, the 470 and 520 nm bleaches are almost equal in the 227 and 766 ps DADS. The shapes of the final DADS and of the final EADS are (by definition) identical.

Surprisingly, the transient absorption data show no indication of EET from carotenoid to porphyrin. It was confirmed by fluorescence excitation experiments performed by Pillai et al. (2013) that EET from carotenoid to porphyrin is indeed absent.

The kinetic scheme that was used in our target analysis with the estimated microscopic rate constants is given in Fig. 6. The carotenoid is excited to the hot S_2 state, followed by relaxation to the S_2 state, where internal conversion to the hot S_1 state competes with ET to the porphyrin. The

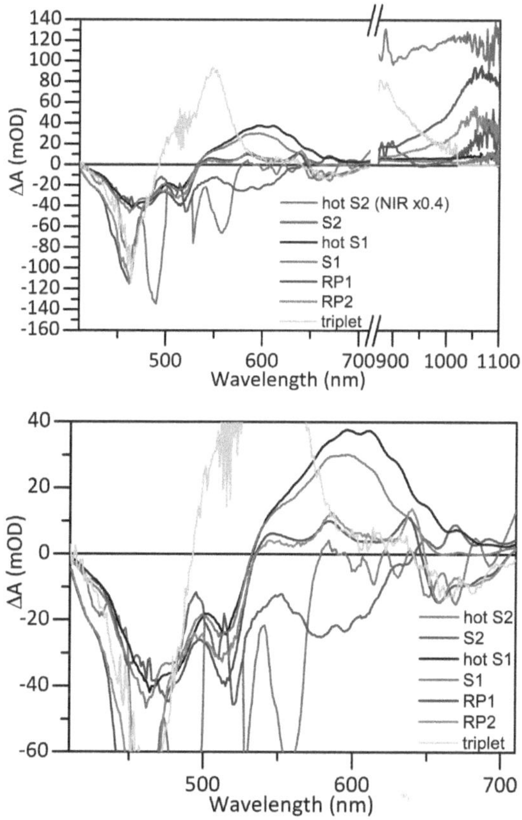

Figure 7 SADS of dyad 1 in toluene using the kinetic scheme of Fig. 6 (top) and zoom of the visible region (bottom). *RP*, Radical pair.

populated hot S_1 state relaxes to the S_1 state. Both the hot and relaxed S_1 state possess identical rates for ET to the porphyrin and for decay to the ground state by internal conversion. The biexponential decay of the radical pairs was modelled with a sequential scheme, which simplifies the charge separation description to single rates from each Car excited state. The adequacy of this kinetic scheme crucially depends upon the physicochemical interpretability of the estimated SADS. These are depicted in Fig. 7. In the zoom of the visible region all expected features are present. The last three states, RP1, RP2 and triplet show a bleach in the porphyrin Soret and Q regions (460 and 650–700 nm). In addition, RP1 and RP2 show three ESA bands between 530 and 630 nm. The carotenoid bleach at 515 nm is smaller with the S_2, hot S_1 and S_1 states than with RP1 and RP2, because ESA compensates this bleach. In the NIR we find almost flat ESA with the hot S_1 and S_1 states. The 1060 nm bands are entirely encompassed in the RP1 and RP2 SADS. Why this band is twice bigger in RP1 remains to be explained. The model carotenoid has been measured by Berera, Herrero, et al. (2006) and Berera, Moore, et al. (2006), who reported an S_1 decay rate of 0.13/ps in toluene, ascribed to IC to the ground state. This agrees well with the 0.125/ps rate estimated. The estimated ET rates are 0.08/ps from the hot S_1 and S_1 states, and 1/ps from the S_2 state. The total RP yield is thus 46%, to which the S_2, hot S_1 and S_1 contribute, respectively, 12%, 11% and 23%. The total triplet yield is 3.8%, which means that 50% of the excitations decay to the ground state from the S_2, hot S_1 and S_1 states.

Summarizing, careful global analysis taking into account the wavelength dependence of the IRF resolved an additional state, the carotenoid hot S_2. Target analysis (Fig. 6) quantified the ET pathways: 0.08/ps from the hot S_1 and S_1 states and 1/ps from the S_2 state. All estimated SADS (Fig. 7) were well interpretable.

3.2 Dyad 2

Dyad **2** (Fig. 2) has been excited in toluene solution in its carotenoid moiety at ≈470 nm. The coherent artefact contributions are so large that we exclude the 466–472 nm traces from the analysis. At least six components are needed for a sufficient fit of the data. With six components (lifetimes 66 fs, 0.59 ps, 6.6 ps, 196 ps, 730 ps, long lived) the overall rms error of the fit is 0.242, but a trend is visible in the first left singular vector of the residual matrix of the visible data. Addition of a seventh component (lifetimes 86 fs, 226 fs, 2.0 ps, 12.5 ps, 143 ps, 514 ps, long lived) results in a significant decrease of the overall rms error of the fit to 0.222, and the above-mentioned trend is no longer visible. This is demonstrated by a comparison of the first left singular vectors of the residual matrix of the visible data. The $u_1^{res}(t)$ of the model with six lifetimes shows clear trends between 0 and 500 ps, which are absent in the model with seven lifetimes, cf. Fig. 8A and B.

The EADS of the seven component models are presented in Fig. 9. The first EADS in global analysis corresponds to the carotenoid S_2 spectrum, of which the lifetime was 86 fs (Fig. 9, grey). The second EADS (green, 226 fs) consists of carotenoid ground state bleach, carotenoid hot S_1 excited state absorption from 520 to 700 nm and carotenoid radical cation absorption at 900 nm. The presence of the signature of both S_1 and the charge-separated state in the second EADS implies that the carotenoid hot S_1 state

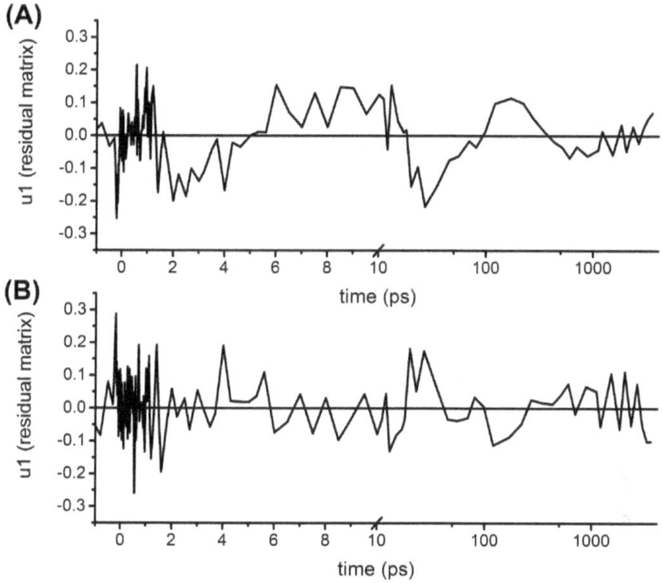

Figure 8 First left singular vector of the residual matrix of the visible data of dyad 2 in toluene after 473 nm excitation using a model with six (A) or seven (B) lifetimes. Note that the time axis is linear until 10 ps (after the maximum of the instrument response function), and logarithmic thereafter.

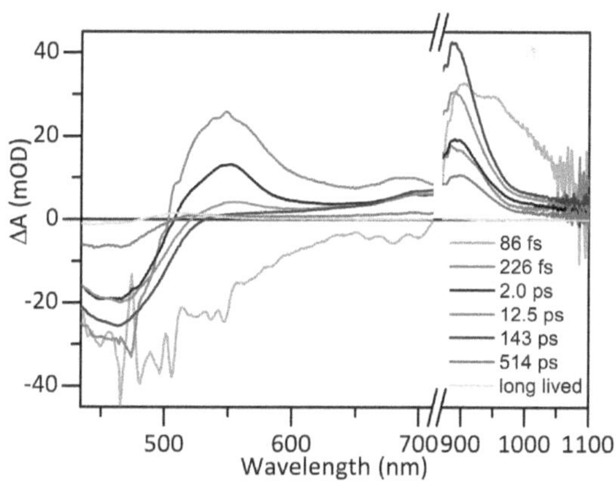

Figure 9 EADS of dyad 2 in toluene after 473 nm excitation.

decays via at least two pathways, IC to S_1 and ET to the fullerene. EET from S_2 to the fullerene could also take place, but it cannot be identified because the spectral signature of singlet-excited fullerene is absent in the probed spectral window, since there is no change in the carotenoid bleach at

450 nm. Due to the shorter conjugation length of this carotenoid compared to the one in dyad **1**, the carotenoid signals are shifted to the higher energy side of the spectrum. The anion of pyrollidinofullerene that was previously reported around 1010 nm ($\varepsilon = 8 \times 10^3\ M^{-1}\ cm^{-1}$) (Guldi & Prato, 2000) was not detected and is probably obscured by the large signal of the carotenoid radical cation ($\varepsilon = 10^5\ M^{-1}\ cm^{-1}$) (Jeevarajan et al., 2011).

The second EADS (green) evolves in 226 fs into the third EADS (black). This evolution includes a large decay of the S_1 excited state absorption and a decrease of the carotenoid ground state bleach. Note that this is accompanied by only a minor increase of the radical cation band, suggesting that the carotenoid hot S_1 transfers most of its energy to the fullerene. The third EADS (black) evolves in 2.0 ps to the fourth EADS (red). The growth of the radical cation band shows that ET takes place from the carotenoid S_1 state to the fullerene, preserving the carotenoid ground state bleach. Most strikingly, when the fourth EADS (red) evolves in 12.5 ps into the blue EADS, the carotenoid ground state bleach increases again, accompanied by a further increase of the 900 nm radical cation band. This can be interpreted as a charge separation process from the lowest fullerene excited state, whose energy level is estimated at 14,000/cm from the absorption spectrum. Decay of fullerene to the ground state can be neglected, as the lifetime of fullerene and derivatives is reported to be more than a nanosecond (Fujitsuka, Luo, Ito, Murata, & Komatsu, 1999). The last three steps are analogous to those of dyad 1. The radical pair population decays biexponentially with lifetimes of 143 and 514 ps, and a small triplet population remains (cyan). Selected traces and fits are depicted in Fig. 10. The decrease, increase and ultimately, decay of the carotenoid bleach are clearly seen at 435 nm. Decay of the S_2 is visible as a spike at 494, 506, 512, 541 and 1013 nm. The rise and decay of the radical pair are visible in all traces, except for 541 nm. At 541 nm the decay of ESA from the carotenoid (hot) S_1 state dominates. The final triplet state is clearly present at 506 and 512 nm.

The DADS (Fig. 11) show clear rises of the radical cation band at 900 nm with time constants of 2.0 and 12.5 ps (black and red). A minor rise with 226 ps (light green) is also discernible. In the carotenoid bleach region the positive red band indicates the 12.5 ps rise attributed to ET from Ful*, the fullerene excited state. The small band at 550 nm is interpreted as slow ET from a population of carotenoid S_1 taking place on the same time scale.

The kinetic scheme used for the target analysis is more complicated than with dyad 1. We need an additional fullerene excited state, which is *assumed*

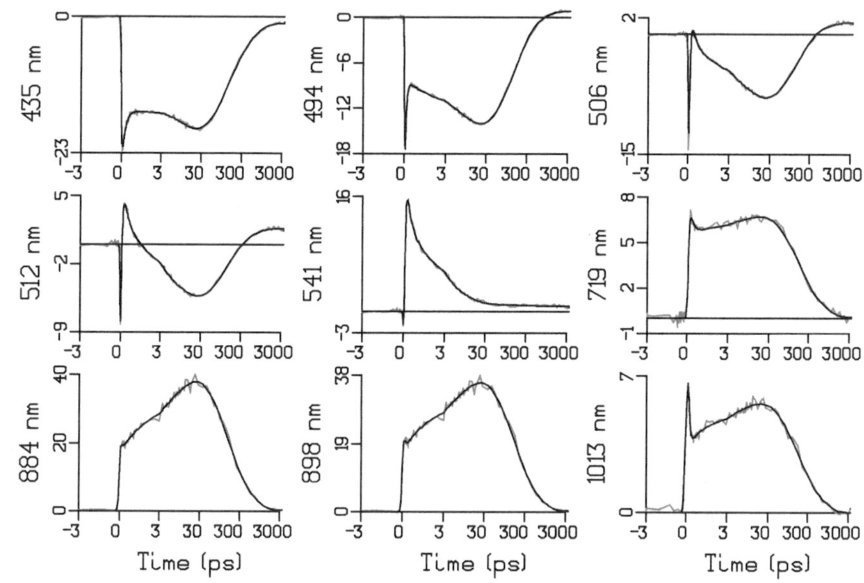

Figure 10 Selected time traces of data (grey, in mOD) and fit (black) of dyad 2 in toluene. Note that the time axis is linear until 3 ps (after the maximum of the instrument response function), and logarithmic thereafter.

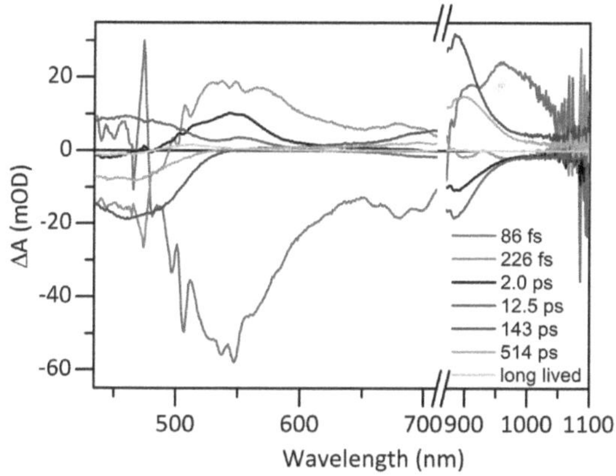

Figure 11 DADS of dyad 2 in toluene.

to possess a zero SADS in the measured wavelength region since there is virtually no bleach nor SE because fullerene has an absorption band around 300 nm (Guldi & Prato, 2000). We will return to this assumption later. In addition a slowly decaying carotenoid S_1 state is present, as can be seen from the 12.5 ps DADS and EADS. Leaving out the hot S_2 state, we

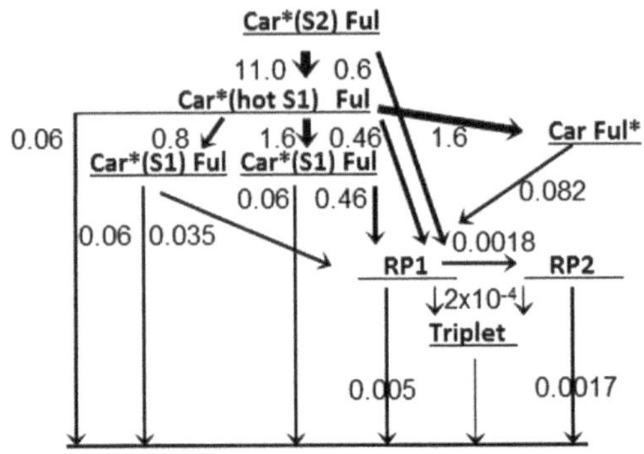

Figure 12 Kinetic scheme for dyad 2 with estimated rate constants in 1/ps. *Car*, Carotenoid; *RP*, radical pair.

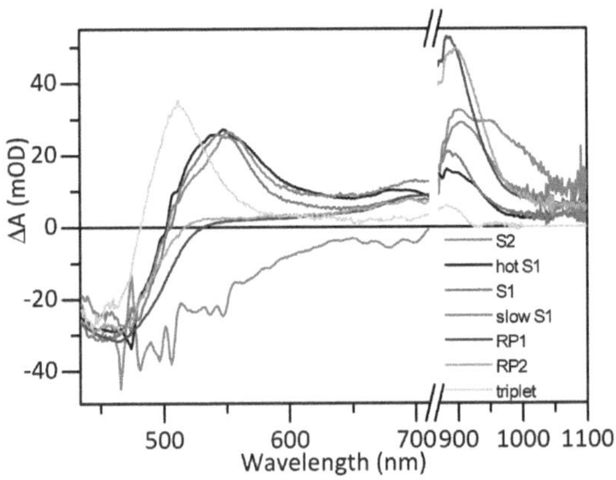

Figure 13 SADS of dyad 2 in toluene using the kinetic scheme of Fig. 12. Note that the Ful* SADS is assumed to be zero in this wavelength region.

need eight species, one more than the number of estimated components in the global analysis. After careful adjustment of the branchings from the carotenoid hot S_1 state, we could estimate the ET rates from the carotenoid (hot) S_1 and Ful★. The ET rates from the carotenoid (hot) S_1 are about six times larger than with dyad 1, 0.46 versus 0.08/ps. This together with the contribution via Ful★ results in a total RP yield of 79%, to which Ful★ and the carotenoid excited states contribute, respectively, 31% and 48% (cf. the dashed and solid RP3 of Fig. 14). The total triplet yield is 4.6%, which means that only 16% of the excitations decay to the ground state from the carotenoid S_2, hot S_1 and S_1 states. The populations of the states are depicted in Fig. 14.

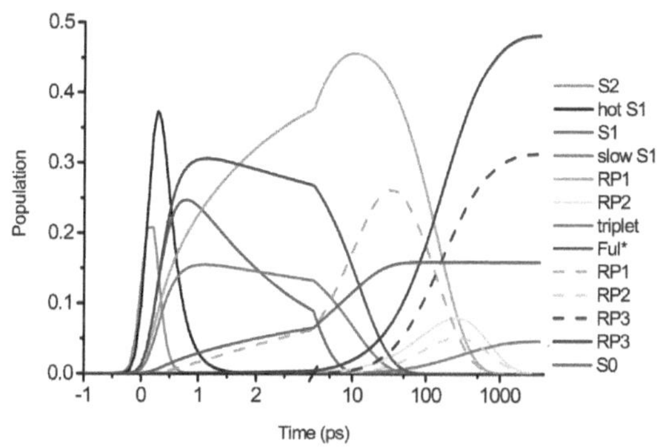

Figure 14 Populations of the states, dashed lines indicate RP (radical pair) created from Ful*. RP3 integrates the decaying RP1 and RP2 states, whereas S_0 integrates the decaying S_2, and (hot) S_1 states.

The estimated SADS are presented in Fig. 13. The three carotenoid hot S_1 and S_1 states (black, red and orange) are very similar: bleach around 450 nm and ESA around 550 nm. Note that this ESA extends to 730 nm in the visible, and in the NIR a peak around 900 nm is present. This peak is the largest with the slowly decaying carotenoid S_1 state (orange). Several explanations are possible. Perhaps a small fraction of the RP decays on this 12.5 ps time scale. More likely is a contribution of the ESA of Ful★ (Guldi & Prato, 2000), which we have assumed to be negligible. Apparently, this assumption is incorrect, and this leads to a reinterpretation of the ESA in the 650−1000 nm region of the (hot) S_1 state. This ESA is now attributed to a mixed excited state (Car-Ful)★. An experiment with fullerene excitation could shed more light on this. Then a simultaneous target analysis of the experiments with carotenoid or fullerene excitation could further clarify the complicated dynamics of the excited state manifold of dyad 2.

Above 1050 nm the signals are more noisy, therefore, we used spectral constraints in that region: the black and red SADS are assumed equal, and the orange SADS has the same shape, but a larger amplitude. The triplet SADS is assumed to be zero above 1000 nm. In this way, four SADS can be resolved above 1050 nm from the more noisy data.

In conclusion, careful global analysis resolved the 2.0 and 12.5 ps lifetimes and the 226 fs lifetime of the carotenoid hot S_1 state. Target analysis (Fig. 12) established that this state decays in five ways: to Ful★ and S_1 in (both with 1.6/ps), to a population of S_1 showing slow ET (with 0.8/ps), to RP1 (with 0.46/ps) and to the ground state (with 0.06/ps). All estimated SADS (Fig. 13) were well interpretable and indicate a mixed excited state (Car-Ful)★.

4. CONCLUSION

In the above described case studies, we have demonstrated that the target analysis methodology is mandatory to describe the processes of energy transfer and charge transfer in artificial photosynthetic systems. Thus we have developed minimal models, identified the steps, and estimated the parameters that characterize the energy conversion in these systems. With this modeling tool the molecular mechanisms of the loss processes have been identified and quantified. This can then inspire photosynthetic device optimization by precisely targeting those sites involved in the most significant losses.

ACKNOWLEDGEMENTS

RvG and IHMvS were supported by ERC Advanced Grant Photprot (to RvG). JJS and JR were supported by the BioSolar Cells Programme of the Dutch Ministry of Economic Affairs. JTMK was supported by a VICI grant of the Chemical Sciences council of the Netherlands Organization of Scientific Research (NWO-CW).

REFERENCES

Alexandre, M. T. A., Luehrs, D. C., van Stokkum, I. H. M., Hiller, R., Groot, M.-L., Kennis, J. T. M., & Van Grondelle, R. (2007). Triplet state dynamics in peridinin-chlorophyll-a-protein: a new pathway of photoprotection in LHCs? *Biophysical Journal, 93*(6), 2118–2128. http://dx.doi.org/10.1529/biophysj.107.106674.

Angerhofer, A., Bornhäuser, F., Gall, A., & Cogdell, R. J. (1995). Optical and optically detected magnetic resonance investigation on purple photosynthetic bacterial antenna complexes. *Chemical Physics, 194*(2–3), 259–274. http://dx.doi.org/10.1016/0301-0104(95)00022-G.

Berera, R., van Grondelle, R., & Kennis, J. T. M. (2009). Ultrafast transient absorption spectroscopy: principles and application to photosynthetic systems. *Photosynthesis Research, 101*(2–3), 105–118.

Berera, R., Herrero, C., van Stokkum, I. H. M., Vengris, M., Kodis, G., Palacios, R. E., … Kennis, J. T. M. (2006). A simple artificial light-harvesting dyad as a model for excess energy dissipation in oxygenic photosynthesis. *Proceedings of the National Academy of Sciences of the United States of America, 103*(14), 5343–5348.

Berera, R., Moore, G. F., van Stokkum, I. H. M., Kodis, G., Liddell, P. A., Gervaldo, M., … Moore, A. L. (2006). Charge separation and energy transfer in a caroteno–C60 dyad: photoinduced electron transfer from the carotenoid excited states. *Photochemical and Photobiological Sciences, 5*(12), 1142–1149. http://dx.doi.org/10.1039/B613971J.

Bonetti, C., Alexandre, M. T. A., van Stokkum, I. H. M., Hiller, R. G., Groot, M. L., Grondelle, R.v., & Kennis, J. T. M. (2010). Identification of excited-state energy transfer and relaxation pathways in the peridinin–chlorophyll complex: an ultrafast mid-infrared study. *Physical Chemistry Chemical Physics, 12*(32), 9256–9266. http://dx.doi.org/10.1039/B923695C.

Fleming, G. R. (1986). *Chemical applications of ultrafast spectroscopy*.

Fujitsuka, M., Luo, C., Ito, O., Murata, Y., & Komatsu, K. (1999). Triplet properties and photoinduced electron-transfer reactions of C_{120}, the [2+2] dimer of fullerene C_{60}. *Journal of Physical Chemistry A, 103*(36), 7155–7160. http://dx.doi.org/10.1021/jp990634w.

Fukuzumi, S. (2006). Bioinspired electron-transfer systems and applications. *Bulletin of the Chemical Society of Japan, 79*(2), 177–195.

Gall, A., Berera, R., Alexandre, M. T. A., Pascal, A. A., Bordes, L., Mendes-Pinto, M. M., ... Robert, B. (2011). Molecular adaptation of photoprotection: triplet states in light-harvesting proteins. *Biophysical Journal, 101*(4), 934–942. http://dx.doi.org/10.1016/j.bpj.2011.05.057.

Guldi, D. M. (2002). Fullerene–porphyrin architectures; photosynthetic antenna and reaction center models. *Chemical Society Reviews, 31*(1), 22–36. http://dx.doi.org/10.1039/B106962B.

Guldi, D. M., & Prato, M. (2000). Excited-state properties of C_{60} fullerene derivatives. *Accounts of Chemical Research, 33*(10), 695–703. http://dx.doi.org/10.1021/ar990144m.

Gust, D., Moore, T. A., & Moore, A. L. (2000). Photochemistry of supramolecular systems containing C_{60}. *Journal of Photochemistry and Photobiology B: Biology, 58*(2–3), 63–71.

Hiroshi, I., Kiyoshi, H., Tsuyoshi, A., Masanori, A., Seiji, T., Tadashi, O., ... Yoshiteru, S. (1996). The small reorganization energy of C_{60} in electron transfer. *Chemical Physics Letters, 263*(3–4), 545–550. http://dx.doi.org/10.1016/S0009-2614(96)01244-4.

Hitchcock, A. P., & Laposa, J. D. (1975). Vibrational frequencies of toluene-d5. *Journal of Molecular Spectroscopy, 54*(2), 223–230. http://dx.doi.org/10.1016/0022-2852(75)90076-4.

Holzwarth, A. R. (1995). Time-resolved fluorescence spectroscopy. *Biochemical Spectroscopy, 246*, 334–362.

Holzwarth, A. R. (1996). Data analysis of time-resolved measurements. In J. Amesz, & A. J. Hoff (Eds.), *Biophysical techniques in photosynthesis* (pp. 75–92). Dordrecht, The Netherlands: Kluwer.

Jeevarajan, J. A., Wei, C. C., Jeevarajan, A. S., & Kispert, L. D. (2011). Optical absorption spectra of dications of carotenoids. *Journal of Physical Chemistry, 100*(14), 5637–5641. http://dx.doi.org/10.1021/jp9530663.

Kloz, M., Pillai, S., Kodis, G., Gust, D., Moore, T. A., Moore, A. L., ... Kennis, J. T. M. (2011). Carotenoid photoprotection in artificial photosynthetic antennas. *Journal of the American Chemical Society, 133*(18), 7007–7015. http://dx.doi.org/10.1021/ja1103553.

Kloz, M., Pillai, S., Kodis, G., Gust, D., Moore, T. A., Moore, A. L., ... Kennis, J. T. M. (2012). New light-harvesting roles of hot and forbidden carotenoid states in artificial photosynthetic constructs. *Chemical Science, 3*(6), 2052–2061. http://dx.doi.org/10.1039/c2sc01023b.

Kodis, G., Herrero, C., Palacios, R., Mariño-Ochoa, E., Gould, S., de la Garza, L., ... Kennis, J. T. M. (2004). Light harvesting and photoprotective functions of carotenoids in compact artificial photosynthetic antenna designs. *Journal of Physical Chemistry B, 108*(1), 414–425. http://dx.doi.org/10.1021/jp036139o.

Koyama, Y., Kakitani, Y., Miki, T., Christiana, R., & Nagae, H. (2010). Excited-state dynamics of overlapped optically-allowed $1B_u^+$ and optically-forbidden $1B_u^-$ or $3A_g^-$ vibronic levels of carotenoids: possible roles in the light-harvesting function. *International Journal of Molecular Sciences, 11*(4), 1888–1929. http://dx.doi.org/10.3390/ijms11041888.

Marcus, R. A., & Sutin, N. (1985). Electron transfers in chemistry and biology. *Biochimica et Biophysica Acta (BBA) – Reviews on Bioenergetics, 811*(3), 265–322. http://dx.doi.org/10.1016/0304-4173(85)90014-X.

Mullen, K. M., & van Stokkum, I. H. M. (2007). TIMP: an R package for modeling multiway spectroscopic measurements. *Journal of Statistical Software, 18*(3).

Nagle, J. F., Parodi, L. A., & Lozier, R. H. (1982). Procedure for testing kinetic-models of the photocycle of bacteriorhodopsin. *Biophysical Journal, 38*(2), 161–174.

Palacios, R. E., Kodis, G., Herrero, C., Ochoa, E. M., Gervaldo, M., Gould, S. L., ... Moore, A. L. (2006). Tetrapyrrole singlet excited state quenching by carotenoids in an artificial photosynthetic antenna. *Journal of Physical Chemistry B, 110*(50), 25411–25420. http://dx.doi.org/10.1021/jp0645317.

Pillai, S., Ravensbergen, J., Antoniuk-Pablant, A., Sherman, B. D., van Grondelle, R., Frese, R. N., ... Kennis, J. T. M. (2013). Carotenoids as electron or excited-state energy donors in artificial photosynthesis: an ultrafast investigation of a carotenoporphyrin and a carotenofullerene dyad. *Physical Chemistry Chemical Physics, 15*(13), 4775–4784. http://dx.doi.org/10.1039/c3cp50364j.

Polívka, T., & Sundstrom, V. (2004). Ultrafast dynamics of carotenoid excited states: from solution to natural and artificial systems. *Chemical Reviews, 104*(4), 2021–2072.

R Development Core Team. (2010). *R: A language and environment for statistical computing.* Vienna, Austria: R Foundation for Statistical Computing.

Snellenburg, J. J., Laptenok, S. P., Seger, R., Mullen, K. M., & van Stokkum, I. H. M. (2012). Glotaran: a Java-based graphical user interface for the R-package TIMP. *Journal of Statistical Software, 49*(3), 1–22. http://www.jstatsoft.org/v49/i03/.

van Stokkum, I. H. M., & Bal, H. E. (2006). A problem solving environment for interactive modelling of multiway data. *Concurrency and Computation: Practice and Experience, 18*, 263–269. http://dx.doi.org/10.1002/cpe.916.

van Stokkum, I. H. M., Larsen, D. S., & van Grondelle, R. (2004). Global and target analysis of time-resolved spectra. *Biochimica et Biophysica Acta, 1657*, 82–104.

van Stokkum, I. H. M., van Oort, B., van Mourik, F., Gobets, B., & van Amerongen, H. (2008). (Sub)-picosecond spectral evolution of fluorescence studied with a synchroscan streak-camera system and target analysis. In T. J. Aartsma, & J. Matysik (Eds.), *Biophysical techniques in photosynthesis* (Vol. II, pp. 223–240). Dordrecht, The Netherlands: Springer.

Williams, R. M., Vân Anh, N., & van Stokkum, I. H. M. (2013). Triplet formation by charge recombination in thin film blends of perylene red and pyrene: developing a target model for the photophysics of organic photovoltaic materials. *Journal of Physical Chemistry B, 117*(38), 11239–11248. http://dx.doi.org/10.1021/jp402086p.

Williams, R. M., Zwier, J. M., & Verhoeven, J. W. (1995). Photoinduced intramolecular electron-transfer in a bridged C-60 (acceptor) aniline (donor) system photophysical properties of the first active fullerene diad. *Journal of the American Chemical Society, 117*(14), 4093–4099. http://dx.doi.org/10.1021/ja00119a025.

Wróbel, D., & Graja, A. (2011). Photoinduced electron transfer processes in fullerene–organic chromophore systems. *Coordination Chemistry Reviews, 255*(21–22), 2555–2577. http://dx.doi.org/10.1016/j.ccr.2010.12.026.

Zhang, J. P., Inaba, T., Watanabe, Y., & Koyama, Y. (2000a). Excited-state dynamics among the $1B_u^+$, $1B_u^-$ and $2A_g^-$ states of all-trans-neurosporene as revealed by near-infrared time-resolved absorption spectroscopy. *Chemical Physics Letters, 332*(3–4), 351–358. http://dx.doi.org/10.1016/s0009-2614(00)01275-6.

Zhang, J. P., Inaba, T., Watanabe, Y., & Koyama, Y. (2000b). Sub-picosecond time-resolved absorption spectroscopy of all-trans-neurosporene in solution and bound to the LH2 complex from *Rhodobacter sphaeroides* G1C. *Chemical Physics Letters, 331*(2–4), 154–162. http://dx.doi.org/10.1016/s0009-2614(00)01165-9.

Zigmantas, D., Polívka, T., Hiller, R. G., Yartsev, A., & Sundström, V. (2001). Spectroscopic and dynamic properties of the peridinin lowest singlet excited states. *Journal of Physical Chemistry A, 105*(45), 10296–10306.

CHAPTER SEVEN

European and International Initiatives in the Field of Artificial Photosynthesis

V. Artero[*,§,¶,1], F. Chandezon[*,§,¶,1], D.T. Co[||], B. Dietzek[#]

[*]University of Grenoble Alpes, Grenoble, France
[§]CNRS, LCBM and SPrAM, Grenoble, France
[¶]CEA, BIG/CBM and INAC-SPrAM, Grenoble, France
[||]Northwestern University, Evanston, IL, United States
[#]Leibniz Institute of Photonic Technology, Jena, Germany
[1]Corresponding authors: E-mail: vincent.artero@cea.fr; frederic.chandezon@cea.fr

Contents

Abstract

In a context of the rising awareness of the need for alternative renewable energy sources to achieve a low-carbon society, artificial photosynthesis, that is, the possibility to directly convert solar energy to chemical energy, is an appealing emerging energy technology. Mimicking the photosynthetic process, the endeavour is to produce a fuel, called solar fuel. We present here a bibliometric analysis showing the evolution of the field in recent years and mapping the emerging academic key actors. This young

Advances in Botanical Research, Volume 79
ISSN 0065-2296
http://dx.doi.org/10.1016/bs.abr.2016.03.002

multidisciplinary research area is swiftly developing, and several European and international initiatives have emerged in recent years to foster its development towards the market. Although artificial photosynthesis is still at the laboratory level, it is also attracting the interest of industries. Many national initiatives in Europe, United States and Asia focussing on artificial photosynthesis have been launched. On the European side, the field begins to be addressed in the agendas of the EU framework programmes as well as in dedicated COST actions (Perspect-H_2O) or in the Joint Programme 'Advanced Materials and Processes for Energy Applications' from the European Energy Research Alliance. Other international initiatives focussing on artificial photosynthesis are the International Energy Agency, the Solar Fuels Institute and the Energy Innovation Hub and several Energy Frontier Research Centers in United States. There are also dedicated conferences such as the Gordon Research Conference series on Solar Fuels and the recently launched International Solar Fuels conference. It is expected that as the field develops, new initiatives and new actors will emerge in different parts of the world.

1. INTRODUCTION

On the path to an energy transition away from fossil fuels and towards sustainable sources, the European Union (EU) is for the moment keeping pace with its objectives, as outlined in the Strategic Energy Technology Plan (SET Plan). By the year 2020, the EU aims at having a 20% share of renewable energy sources (RES) in its energy panel, this number being an average target for the 28 member countries of the Union and the target varying from country to country (Thapper et al., 2013). With a current growth of nearly 7% between years 2004 and 2013 to reach an RES of 14.95% in 2013, the objective of 20% should be easily met and even surpassed, provided that the growth follows the same trend (Eurostat shares, 2013). Looking beyond 2020, the objectives are even more ambitious with a targeted share of RES in the global energy panel of more than 50% and a concomitant reduction of greenhouse gas emissions of more than 80% (Energy Roadmap 2050, 2013). Outside Europe, due to the raising awareness of the dramatic impact that climate change could have on humanity, there are also similar initiatives to increase the share of low carbon energy sources and reduce drastically the emissions of greenhouse gases. The US Department of Energy has continued their commitment to solar and renewable energy research through the renewals of the Energy Frontier Research Center (EFRC, see Section 3.2.3 for more details below) programme and the Joint Center for Artificial Photosynthesis (JCAP) Energy Innovation Hub (JCAP, 2010).

The current energy technologies would not allow meeting these ambitious targets. Therefore for keeping a steady and significant growth of the share of RES beyond 2020, scientific breakthroughs must be achieved in the future. The sun is an almost unlimited source of renewable energy: in 1 h the solar radiation energy that falls on earth is higher than the total energy consumption of the world in one year (Lewis and Nocera, 2006). One method for harvesting solar energy is the photovoltaic effect that converts solar energy to electricity. Yet, as this is an intermittent source of energy, also unevenly distributed on the planet, solar energy has to be stored for providing power during night time and cloudy days.

A more advantageous solution would be to directly convert solar energy to a chemical form, a fuel, that can be stored and transported for later use, for example for generating electricity or for transportation, taking into account that more than 80% of the primary energy is based on fuels (Thapper et al., 2013). This is a grand scientific challenge. One important approach to achieve this goal is artificial photosynthesis, a discipline which nowadays gathers materials science (Photo—Electro—Chemical processes), molecular sciences (antenna systems, light-driven electron transfers and bioinspired catalysts) and biological sciences (photosynthesis and multielectron metalloenzymes) through the use of common concepts (light harvesting, electron transfer, electrocatalysis, reaction mechanisms) and tools (nanosciences, electrochemistry, photochemistry, advanced spectroscopy and theoretical chemistry). Fig. 1 displays a word cloud from the most cited keywords found during a bibliometric survey that is discussed in the following sections. It

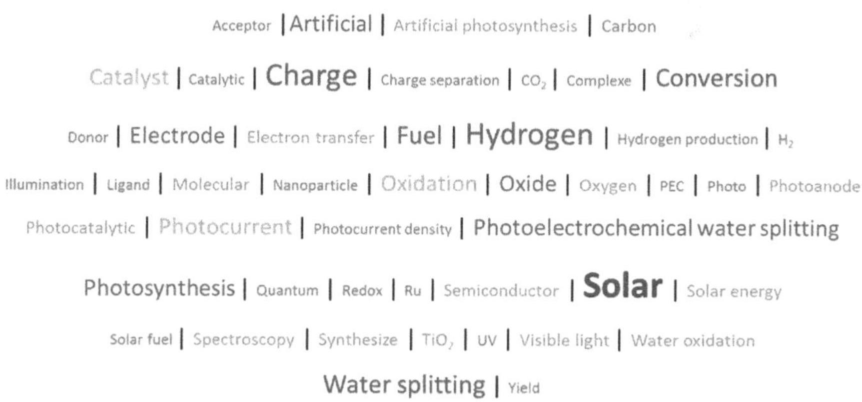

Figure 1 Keyword cloud illustrating the field of solar fuels, artificial photosynthesis and photoelectrochemical catalytic processes. These keywords are the most cited concepts extracted from titles and abstracts of the 4298 articles found during a bibliometric analysis made using Thompson Reuters Web of Science database (INTELLIXIR). See the text for further details on the search procedure.

clearly shows the multiple interfaces among biology, chemistry and physics within this recent field.

This chapter aims at presenting this young yet swiftly developing multidisciplinary research area and the related European and international initiatives that have emerged in recent years to foster it. In the following section, after a brief discussion on the definition of artificial photosynthesis, we present and discuss a bibliometric analysis to show the evolution of the field in the recent years and the emerging key actors. We also discuss the conferences dedicated to this research area as well as present the industrial context. The ensuing section will be dedicated to presenting collaborative initiatives, for example in the frame of the EU framework programmes as well as national and international networks and organizations.

2. ARTIFICIAL PHOTOSYNTHESIS: ACADEMIC AND INDUSTRIAL CONTEXT

2.1 Academic Research on Artificial Photosynthesis

To get a clear and unbiased idea of the field and its recent evolution, we performed a bibliometric analysis using three sets of keywords. The first set combined the keyword 'photoelectrochemical' with one from the following list: 'water splitting', 'hydrogen' or 'CO_2 reduction'. The second set included the words 'solar' and 'fuels'. Finally, the third set focussed on 'photosynthesis' combined with either 'mimic' or 'artificial'. We note that the definition of the field with this limited set of keywords excludes subfields such as biochemistry and bioinspired chemistry, when not explicitly coupled to light-driven processes, and photosynthetic research, when not interfaced with other fields. Interrogating the Thompson Reuters Web of Science database (Web of Science), we found 4298 articles from 9526 authors belonging to 410 affiliations from 74 countries between 1957 and 2014. These data were analysed using the software INTELLIXIR©. Fig. 2 (top panel) clearly shows that the field has grown significantly recently. Only few, though seminal, papers were published before the 1990s, likely a response to the second oil crisis in 1978. The field now grows exponentially since the year 2000. This threshold corresponds to the rise of the occurrence of the three exact keywords 'artificial photosynthesis', 'solar fuels' and 'photoelectrochemical water splitting' in our bibliometric survey. Although the sum of occurrence of all three keywords only accounts for half of the yearly published papers, this clearly indicates that the community has begun to

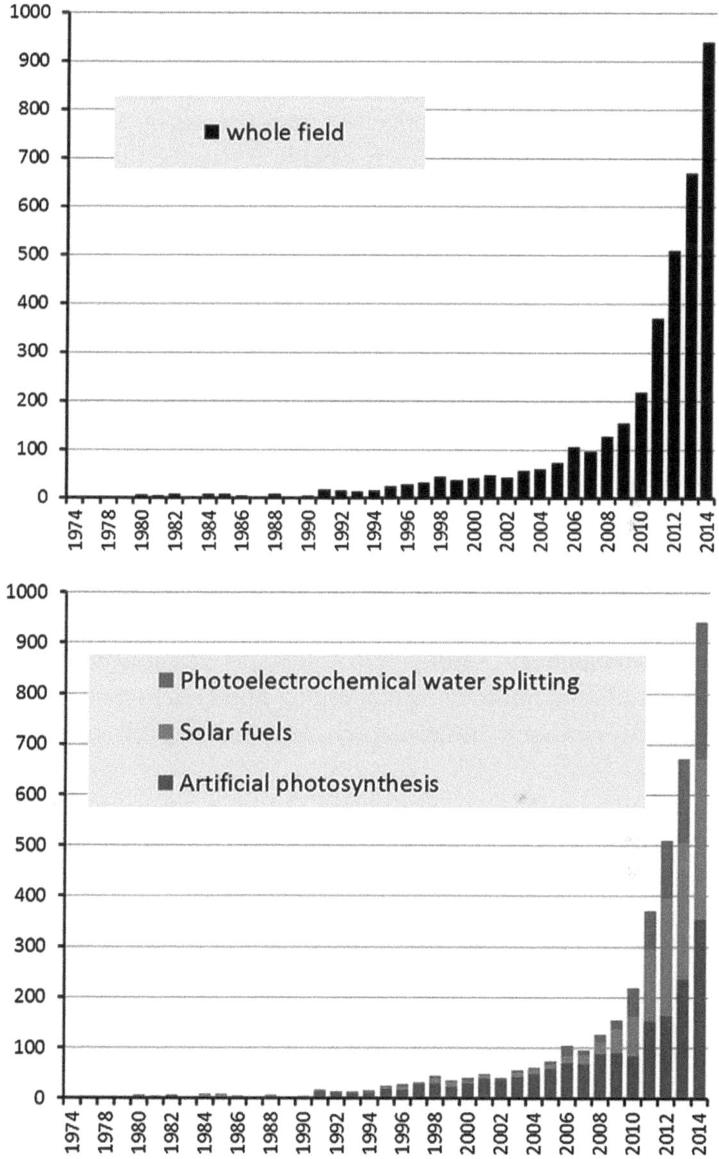

Figure 2 Bibliometric survey of the field of artificial photosynthesis using three sets of keywords and the Thompson Reuters Web of Science database. See the text for further details on the search procedure. Top panel: Global repartition per year of the resulting publications. Bottom panel: Repartition per year for each set of keywords.

organize itself through the use of well-defined tags to be more visible and to efficiently target stakeholders and funding agencies.

A detailed analysis of the three independent sets (Fig. 2, bottom panel) reveals that there is almost no overlap between the use of 'solar fuels', 'artificial photosynthesis' and 'photoelectrochemical' (applied to catalytic reactions such as water splitting and CO_2 reduction) as keywords to designate the field. 'Artificial photosynthesis' has been used from the premises of the field. The term 'solar fuels', coined by Bolton in 1978, has not been significantly used before the emergence of the Gordon Research Conference on 'Renewable energy: Solar fuels' in 2007, but has been adopted by the community since that time. Finally, the subfield related to 'photoelectrochemical water splitting and CO_2 reduction processes' also has only significantly grown since 2010, in line with the report of a variety of fully operating systems in the recent years, and this indicates that the field now has become mature enough to qualify the reports with technologically relevant keywords (Ager, Shaner, Walczak, Sharp, & Ardo, 2015). In addition, such an absence of overlap in the keywords selection by authors indicates the co-existence of three communities sharing similar goal and concepts but coming from various disciplines: chemical physicists and physical chemists rather qualify their studies as photoelectrochemical. The same is true for materials scientists while synthetic molecular chemists prefer the term 'artificial photosynthesis'.

At the international level, pioneers of the field are M. Grätzel (EPFL, Lausanne Switzerland), T. J. Meyer (Univ. North Carolina, Chapel Hill, USA), B. A. Parkinson (Solar Energy Research Institute, now NREL, Golden, CO, USA) and T. Moore, A. Moore and D. Gust (Arizona State University, USA). The group of M. R. Wasielewski (Northwestern University, USA) and the Swedish Consortium for Artificial Photosynthesis, gathering eminent scientists such as S. Styring, L. Hammarström, V. Sundström and L. C. Sun came in the 90s, concomitantly to the emergence of efficient time-resolved spectroscopic methods and important discoveries in structural biology with the identification of three-dimensional structures of the active sites of hydrogenase enzymes and photosystems.

The current top prominent institutions active in the field as identified from our bibliometric survey include the University of California, Berkeley, University of North Carolina at Chapel Hill, Caltech (CA), Stanford (CA), Arizona State University, Penn State, University of Texas, Northwestern University (IL), Yale (CT) together with Argonne National Laboratory and the National Renewable Energy Laboratory in the United States. In

Asia, the main players are the Chinese Academy of Sciences (China), the Chinese universities of Beijing, Jiangsu, Dalian, Tianjin and Shanghai, the University of Tokyo, Osaka and Tohoku (Japan), the National University of Singapore and Nanyang Technological University (Singapore), the University of Seoul and Korean Institute of Science and Technology (Korea) and Taipei University (Taiwan). Finally, the list includes in Europe the University of Uppsala, EPFL (Lausanne, Switzerland), the Max Plank Institute for Chemical Energy Conversion and University Erlangen (Germany), Imperial College and University College London (UK) and CNRS (France). However, such a short list does not fully reflect the research effort per country. Fig. 3 shows that Italy, Spain and the Netherlands are also quite active in Europe. Interestingly, countries within the European Union's research network almost equal the USA in terms of publications over the whole period. Similarly the research effort from Asia forms another third of the whole contribution.

Our main motivation for carrying out this bibliometric survey was the quantification of international collaboration in the field. Surprisingly, we found it quite limited, if double affiliations for single authors are discarded. Indeed, some senior scientists from Europe, Japan or the USA have a joint

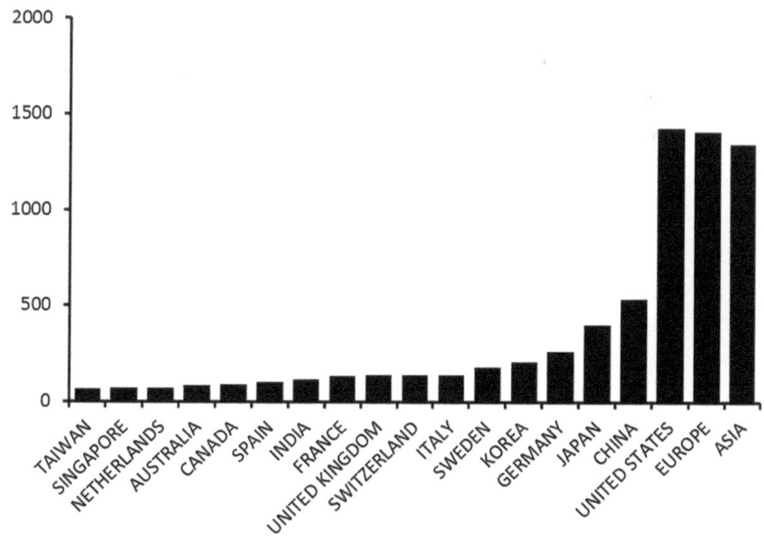

Figure 3 Repartition of research efforts per country in terms of total publications in the field. (Only countries with more than 75 publications listed in the bibliometric survey are listed.) We sum under 'Europe' the contribution of all countries associated in the European research framework.

affiliation with a university in either Singapore, Korea or China, although this does not truly correspond to collaboration between scientists from both countries. Genuine cases of international scientific collaborations are thus rare and mostly correspond to specific collaborations between principal collaborators, with distinct expertise, for example theoretical chemistry in one country and experimental studies in another. We were unable to identify strong international collaborative institutional networks except for the joint centre between Dalian University and Swedish Institutions and the connection between EPFL (Switzerland) and US universities through the NSF Center for Chemical Innovation in Solar Fuels (CCI Solar). Most collaborative publications are found within institutions from the same country as they mirror funding schemes. Surprisingly, the European Research framework at play since 1994 did not place Europe at the top of the list in our survey in terms of international publications and few joint publications, mostly involving the Swedish Consortium for Artificial Photosynthesis, have been found within Europe in this survey relative to the total number of publications considered.

2.2 Conferences

Regarding international conferences dedicated to the field of artificial photosynthesis and solar fuels, to date there are two prominent initiatives. These two international events are coupled to meetings dedicated to young investigators, showing the willingness of the community to prepare and nourish the future of the field.

The oldest one is the 'Renewable Energy: Solar Fuels', Gordon Research Conference (GRC Solar fuels, 2007). It was initiated in 2007 by Daniel G. Nocera (formerly at MIT, now at Harvard University) and takes place every 2 years in alternation between the United States of America and Europe. The organization of sessions in Asia is recurrently discussed. This large Gordon Research Conference (~200 attendees) is chaired by two specialists covering the biological, molecular and materials scientific facets of the field and belonging to distinct continents (typically European and American). The next session of this GRC is planned in early 2018, in California.

A younger yet larger conference is the International Solar Fuels (ISF) conference. It was organized for the first time in April 2015 in Uppsala (Sweden) and gathered approximately 500 delegates, a very high number for a newly initiated conference which emphasizes the rising interest for solar fuels in the world (ISF 1, 2015). The ISF conferences will be organized biannually, the next edition being planned in San Diego (California) in 2017.

A series of meetings on 'Solar Cells and Solar Fuels' are also organized in Dalian (China) every 2 years by L. C. Sun (Dalian Univ. of Technology) and Can Li (Dalian Institute of Chemical Physics). Besides these, other meetings have been organized in the last few years throughout the world. They are usually meetings organized in the frame of established networks in the field of solar fuels and artificial photosynthesis such as the workshop 'What is the efficiency of Photosynthesis?' held in May 2009 in Albuquerque, New Mexico, USA sponsored by the US department of energy (Blankenship et al., 2011), the International Conference on Artificial Photosynthesis (ICARP 2014) organized in November 2014 in Awaji Island (near Osaka) by the Japan network 'AnApple: All Nippon Artificial Photosynthesis Project for Living Earth' (the next meeting is scheduled in 2017) or the UK Solar Fuel Network (SFN) International Discussion: 'Solar Fuels: Moving From Materials To Devices' chaired by J. Barber and H. B. Gray in London in July 2015. Other meetings gather national research communities.

For such a swiftly growing field, it is very likely that new series of conferences will be launched in the future and will spread from North America and Europe where they mostly took place until now — to other regions of the world, such as Asia which proves very active as displayed in Fig. 3.

2.3 Industrial Context

If realized, solar fuels could solve many of the world's energy problems. Yet historically, the technological challenges surrounding solar fuel commercialization have been difficult to overcome. This is not for lack of innovation as already indicated by our bibliometric survey: around the world, prestigious universities, innovators and entrepreneurs are working hard to crack the barriers to scaled, commercial solar fuel production. In other words, the present positioning of the fields of artificial photosynthesis and solar fuels on the TRL scale (Technology Readiness Level) is typically around 3, with a strong involvement of academic actors and funding coming mostly from public sources. At the European level, an effort to support to increase the TRL level of artificial photosynthesis was done through dedicated calls of the Fuel Cells and Hydrogen Joint Technology Initiative (FCH-JTI). Several successful projects gathering academic and industrial partners were funded through these calls, ArtipHyction and PECDEMO (see Section 3.1.2 and Table 1).

The present effort is put now on translating the concept of artificial leaf and tree into working prototypes that could ultimately be transferred to industry to enter the market. This is illustrated in Fig. 4 with the 'artificial tree' of the Solar Fuel Institute (SOFI), an international consortium of universities,

Table 1 List of Some European Projects in the Field of Solar Fuels and Artificial Photosynthesis

Name of the Project	Running Period	Coordinator	Number of Partners/ Countries	Framework Programme
TMR Network CT96-0031: Ru-Mn for artificial photosynthesis	1996→2000	Uppsala University	10/5	FP4
SOLAR-H: Hydrogen from sun and water – Linking two scientific fields	2005→2007	Uppsala University	8/6	FP6
SOLAR-H2: European solar fuel initiative –renewable hydrogen from sun and water. Science linking molecular biomimetics and genetics	2008→2011	Uppsala University	12/8	FP7
SolHydromics: Nanodesigned electrochemical converter of solar energy into hydrogen hosting natural enzymes or their mimics	2009→2012	Politecnico di Torino	7/5	FP7
NanoPEC: Nanostructured photoelectrodes for energy conversion	2009→2012	EPFL	7/7	FP7
ArtipHyction: Fully artificial photoelectrochemical device for low temperature hydrogen production	2012→2015	Politecnico di Torino	8/5	FP7 FCH JTI
PECDEMO: Photoelectrochemical demonstrator device	2014→2017	Helmholtz Zentrum zu Berlin	7/4	FP7 FCH JTI

Figure 4 The SOFI 'artificial tree' concept drawing was crowd-sourced and created by a designer in Serbia. The solar fuels components of photocatalytic leaves, water (blue tanks), CO_2 (air intakes) and liquid fuel (amber tanks) are represented to inspire and mark the beginning of SOFI's Demonstration Project.

government laboratories and industrials which are discussed in detail in Section 3.2.2. SOFI Members, including industry partners, are exploring the assembly of SOFI's first integrated lab-scale system called 'SOFI's Demonstration Project' based upon unique insights gathered through SOFI's processes and diverse membership. This demo project will serve two functions. First, it will illustrate to the public and decision makers that the implementation of liquid solar fuels technology is not farfetched. Second, it will expose integration risks/challenges, identify synergies between siloed technologies/sciences, provide the community with performance data and create a baseline for full-scale design and techno-economics. This 'artificial tree' concept drawing was crowd-sourced and created by a designer in Serbia (see Fig. 4). The solar fuels components of photocatalytic leaves, water (blue tanks), CO_2 (air intakes) and liquid fuel (amber tanks) are represented to inspire SOFI's Demonstration Project.

Going a step further beyond a laboratory prototype, the challenge of cost-effective energy technology scale-up is significant, which is the case for any involving scaled, capital-intensive energy equipment. However, other full applications of the solar to fuels process are being pursued by leading global companies. We could mention in this sense the following two examples:

- German automaker Audi, in collaboration with Climeworks and Sunfire GmbH, has backed a project where CO_2 is extracted from the atmosphere and converted to synthetic diesel-like liquid fuel. The process consists of high-temperature water electrolysis to generate hydrogen followed by conversion to synthesis gas and finally long-chain hydrocarbon compounds via Fischer–Tropsch process.
- The SOLAR-JET project, funded by the European Commission CORDIS Program, demonstrated on a laboratory scale a process that combines concentrated sunlight with CO_2 captured from air and H_2O to produce kerosene by coupling a two-step solar thermochemical cycle based on nonstoichiometric ceria redox reactions with the Fischer–Tropsch process.

As current issues and bottlenecks around artificial photosynthesis are addressed by the academic world and feasibility of working prototypes, therefore TRL, increases, it can be expected that the interest of entrepreneurs and industry will rise, thus pushing further artificial photosynthesis towards the market and helping to turn solar fuels into an economical reality.

3. EUROPEAN AND INTERNATIONAL INITIATIVES

In the previous section, we discussed how the research effort is shared throughout the world as well as the landscape of key players at present. The spectacular emergence of a research community focussed on solar fuels and artificial photosynthesis in the past 20 years led to different initiatives in order to coordinate the research efforts, be it at the national level or through international networks. This is discussed in the present section.

3.1 Europe

3.1.1 National Initiatives

In Europe, the first national network dedicated to solar fuels and artificial photosynthesis was the Swedish Consortium for Artificial Photosynthesis (CAP), set in 1994 as a bottom-up initiative (Swedish Consortium, 1994). This Swedish network gathers university scientists with broad multidisciplinary expertise from molecular biology, biophysics and biochemistry to organometallic and physical chemistry. The main groups active in the consortium are located at the Universities of Uppsala and Lund as well as at the Royal Institute of Technology (KTH) in Stockholm.

Following this pioneering initiative, many other national networks and initiatives emerged in other countries of Europe in the past years.

In the United Kingdom, the SolarCAP Consortium for Artificial Photosynthesis was launched in 2009 for a duration of 3 years with supporting funds from the British Engineering and Physical Sciences Research Council (ESPRC) (SolarCap, 2009). It was also an academic network gathering groups from the Universities of York, East Anglia, Manchester and Nottingham. SolarCap aimed at developing materials and devices for the production of carbon-based solar fuels. A presently active UK network is the Solar Fuels Network (SFN) launched in 2012. It aims at bringing together the UK-based research community in the field of artificial photosynthesis as well as international groups through the organization of dedicated events and the funding of mobility. One such initiative is the UK Solar Fuels symposia initiated in 2012 in London, the fourth edition being planned in January 2016 in Cambridge.

In the Netherlands, BioSolarCells (2009) is a consortium gathering nine Dutch research organizations (including the University of Amsterdam, Leiden and Twente and the Technological Universities of Delft and Eindhoven) together with about 30 industrial partners. It benefits from a grant of 42 M€ from public (Dutch government) and private sources to fund research projects with the aim of building and strengthening links with industry and developing new business opportunities.

In Germany, two nationally funded institutes address solar fuels and artificial photosynthesis. One belongs to the Max Planck Society: MPI for Chemical Energy Conversion, founded in 2012 (formerly the Max Planck Institute of Inorganic Biochemistry in Mülheim an der Ruhr) (MPI CEC, 2012). Besides, the Helmholtz Association of German Research Centres launched in 2008 the Institute of Solar Fuels in Berlin (Helmholtz Institute Solar Fuels). In addition, solar fuels-oriented projects are supported by the German Research Association (Deutsche Forschungsgemeinschaft, DFG) and by the Ministry of Education, Research and Technology (*Bundesministerium für Bildung, Forschung und Technologie*; BMFT) through different programmes. Furthermore, artificial photosynthesis plays a prominent role in the Leibniz Institute for Catalysis (LIKAT) in Rostock.

In Italy, SOLAR-CHEM, the Italian interuniversity centre for the chemical conversion of solar energy was founded in 2009 by the Universities of Bologna, Ferrara and Messina joined later on by groups from other

Universities (Campagna, Bonchio, Venturi, & Scandola, 2012). The main objective of this network is to foster research in solar fuels in Italy through a multidisciplinary approach and by coordination activities, for example through the organization of dedicated events and through short-term exchanges of staff in the network.

In Switzerland, prominent actions are the PECHouse (2011–2014) and the currently running PECHouse2 (Photoelectrochemistry competence centre) initiatives coordinated by Prof. Grätzel (Ecole Polytechnique Fédérale de Lausanne) and involving University of Basel and EMPA (PECHouse 2, 2015). These centres aim at developing both materials and technologies related to solar fuels in Switzerland.

In France, the structuration of the research community in the field of solar fuels and artificial photosynthesis is more recent. A research cluster (*Groupement de Recherches*) entitled SFN-France (Solar Fuel Network France) has been launched under the patronage of CNRS in January 2016 for a duration of 4 years (GDR Solar Fuels, 2016). It will gather more than 35 different teams from all over the country with different scientific backgrounds, again reflecting the multidisciplinarity of the area. This network aims at gathering the research community around solar fuels in France as well as coordinating the research effort. In addition at least two laboratories of excellence (LabEx) funded at the multimillion euro level by the French government over 8 years put artificial photosynthesis at the forefront of their research efforts in Grenoble (LabEx Arcane) and Paris-Saclay (LabEx Charmatt).

In other countries of Europe, there also exist similar national initiatives such as those in Ireland and Spain, and others will certainly emerge in the future. This brief (but nonexhaustive) overview of national initiatives in Europe again illustrates the swiftly rising interest of the academic research community on the topic of solar fuels, in line with the bibliometric analysis discussed earlier. Research communities from different scientific horizons are organizing themselves and cross-fertilizing around the area with the aims (1) to coordinate their research effort in a period where shortage of funds for the academic community is becoming a major concern, (2) to give a better visibility to the area and (3) to build links with industrial partners.

3.1.2 European Projects

European projects can be coined as international initiatives at the European scale. In line with the emergence of the topic of solar fuels and of a related multidisciplinary research community, several European projects have thus been supported in the frame of the different framework programmes, see Table 1. As can be seen, most of these projects were supported in the FP6 and afterwards meaning that it needed some delay for the area to appear in the European agendas. The coordinators of these projects are research groups that were pioneers in the field and that were identified in the bibliometric survey discussed earlier (Section 2.1).

Up to now, we identified only one project in the currently running H2020 framework programme, PECDEMO (2015). In 3 years, it aims at developing a prototype with a Solar-to-Hydrogen efficiency of 8%, beyond the current state-of-the-art. This illustrates the trend to move from fundamental processes and materials towards a device, or, in other words, to reach a higher TRL level, typically 4. This is in line with the current trend in H2020 which puts a strong stress on higher TRL levels as compared to the previous FPs. This can be a serious drawback for the solar fuel area since most barriers still concern the design of novel low-cost materials with improved stability and activity. It is also important that in the forthcoming work programmes of H2020, solar fuels appear more clearly identified and not merged with topics such as biofuels or hydrogen, where the technologies currently reach significantly higher TRLs and are close or have reached the market. First indications in the new work programmes for 2016—2017 are encouraging in the sense that solar fuels and artificial photosynthesis are explicitly mentioned.

3.1.3 European Networks

3.1.3.1 COST Perspect-H_2O

The European Program Cooperation in Science and Technology, COST, is the longest-running European framework aiming to support transnational research across Europe. The primary tools of a COST Action are the exchange of young researchers between research group all around Europe to establish long-lasting personal networks and a European research community in the respective fields. In the context of solar fuels, COST is supporting the Action CM1202 Perspect-H_2O — Supramolecular Photocatalytic Water Splitting since 2012 for a duration of 4 years (Perspect-H_2O, 2012).

Scientifically, Perspect-H_2O is set up to establish a guided design of supramolecular photocatalysts for water splitting, and hence the generation of molecular hydrogen as a renewable fuel. The network that by now integrates research groups from 21 EU and neighbouring countries thus combines synthetic chemistry, photophysics and photochemistry, electrochemistry and theory aiming at (1) a detailed molecular mechanistic understanding of photoinduced reaction steps in supramolecular photocatalytic water splitting and (2) development of functional systems. The envisioned approach identifies relations between molecular and electronic structure, photoinduced structural and electronic dynamics, catalytic efficiency and the stability of long-lived intermediates and catalytically active species. Finally, it aims at the design of functional photocatalytic materials based on earth-abundant elements and prototypes of supramolecular photocatalytic water-splitting systems. To address these challenges, Perspect-H_2O has four working groups (WG): **WG1**: Synthesis and photocatalysis; **WG2**: Device integration; **WG3**: Photoinduced dynamics and **WG4**: Structural intermediates and catalytically active species.

A very important instrument of the COST network allows the funding of short-term (less than 3 months) exchanges between partners. Thus, each year, approximately 15 students, postdocs or permanent scientists are visiting other laboratories to complement ongoing studies within specific measurements or calculations or to learn novel techniques. The COST network also organizes dedicated workshops to discuss key aspects of the field. In that prospect, two issues that CM1202 has tackled in the past were associated with establishing protocols for standardizing tests of supramolecular photocatalysts and testing NiO-based photoanode materials.

- The turnover number as a characteristic value to quantify a photocatalytic system's performance is — at least in the field of supramolecular photocatalytic water splitting — highly dependent on the experimental conditions: solvent composition, nature and concentration of the sacrificial electron donor/acceptor, the irradiation intensity and colour, the volume ratio of solution and head space, and so on. Hence, having standardized protocols for testing supramolecular photocatalysts is considered an issue of utmost importance to boost the field. A meeting has been recently hosted by Johannes G. Vos in Dublin to discuss these points with international experts and the publication of the conclusions of the meeting is in progress. It will present the input of the Action for discussions with, for example, the IUPAC to further nurture the development of international standardization in this vivid field.
- Currently, about half-a-dozen groups in Europe are devoting (part) of their research to the synthesis of NiO-based photocathodes for PECs. As by far most of these groups are active participants in CM1202, they joined their expertise and forces to collaboratively evaluate the individually prepared NiO electrodes under identical conditions. During a meeting hosted by Elizabeth Gibson in June 2014 in Nottingham,

experimental measurements have been performed onto a series of distinct NiO materials. This study aimed at establishing differences between the preparation methods in order to define a state-of-the-art performance of the material, which is not biased by different evaluation strategies. The results of the meeting have been submitted for publication (Wood et al., 2016).

These two examples highlight the ambition of the COST Action PERSPECT-H_2O not only to boost collaborative research and to train young researchers but also to define standards and provide jointly accepted protocols for the evaluation of PEC materials and molecular photocatalysts. Thereby, this COST Action is part of shaping a sustainable European research community in the field of solar fuels and artificial photosynthesis.

3.1.3.2 European Energy Research Alliance and AMPEA, the Joint Programme on Advanced Materials and Processes for Energy Applications

In order to support the transition to a low carbon energy society and to meet its objectives it has fixed in terms of the share of RES by 2020 and 2050, the European Union launched in 2008 the Strategic Energy Technology SET Plan (2008). The main objective of the SET Plan is to accelerate the deployment of low carbon energy technologies by implementing coordinated research and innovation policies throughout the whole union and associated countries. The European Energy Research Alliance (EERA) is the public research pillar in the frame of this SET Plan (EERA, 2008). Indeed, EERA is an alliance of more than 150 European public research centres and universities which, through a tightly focussed strategy, aims at accelerating the development and market uptake of key low-carbon technologies in line with the EU and SET Plan objectives. The main objectives of the EERA can be summarized as follows:

- **Coordinate the research effort around low carbon energy technologies in EU:** This is a key issue in order to avoid duplication of research efforts in a context of constrained research budgets, based on the fact that no country can address all research around low carbon energy technologies and has to set its own priorities. This coordination involves that EERA members actively work together in the frame of so-called Joint Programmes (vide infra) and share priority settings and research projects.
- **Align national research**: The previously mentioned coordination of research effort also involves that member states align their research

policies and agendas around low carbon energy along those of EU as discussed with the EERA and the European Commission. Having national energy research alliances as stakeholders is a clear asset for that. Thus besides existing alliances such as ANCRE in France, EERA fostered the creation of new ones in several EU countries including the UK (UKERC), Netherlands (NERA), Spain (ALINNE) and Belgium (BERA).

- **Collaborate with industry**: EERA is the research pillar in the SET Plan, yet its duties also involve working in close cooperation with European industrial platforms so as to foster world-class technology and innovation in Europe's energy sector, while helping to reduce the time it takes for this technology to reach the market.
- **International cooperation**: As the entire world faces the challenges around climate change and the needed transition towards low carbon energy, international cooperation is also one objective of EERA. Thus, EERA Joint Programmes act as points of contact for collaboration outside Europe. EERA representatives regularly represent the EU scientific community in collaboration with the European Commission at events outside Europe (eg, Asia, USA and Brazil). In addition, exporting the developed energy technologies outside the European Union could also benefit the European industry and economy.

The various energy technologies are addressed in EERA through the Joint Programmes (JPs), each having its own partnership with EERA members active on the JP topic. Presently, EERA has launched 17 Joint Programmes, the full list of which can be consulted on the EERA webpage (EERA, 2008). On a schematic point of view, the 17 JPs can be clustered in three main groups. The first, and largest cluster, comprises 14 JPs which address specific energy technologies such as 'Bioenergy', 'Energy storage' or 'Photovoltaic solar energy'. The JP 'Economic, environmental and social impacts (JP e3s)' form a second separate cluster, which addresses the economic, environmental and societal issues necessarily involved in the transition to a low-carbon society. Finally, the Joint Programme 'Advanced Materials and Processes for Energy Applications' (AMPEA) is focussing on basic science issues and materials for low carbon energy technologies, therefore acting at lower TRLs as compared to other technical EERA JPs (AMPEA, 2011).

Indeed, implementing a basic science programme for energy is one of the key roles assigned to EERA by the European Commission in the frame of the European Strategic Energy Technology Plan (SET Plan). Basic science is often presented as crucial in generating the breakthroughs needed to bring new generations of technologies to the market. The aim of AMPEA is thus to foster a truly multidisciplinary approach to develop enabling tools and new concepts for future emerging energy technologies and to harness and integrate materials science and process innovation for high performance sustainable energy technologies, in order to enhance the long-term competitiveness of European industry. AMPEA aims at working in close coordination with other EERA-JPs and with universities to bridge fundamental and applied research fields. AMPEA presently gathers approximately 34 public research organizations and universities from 14 European countries.

In order to fulfil its missions, AMPEA is organized in a matricial way with the aim of applying the capacity of 3 '*Tools*' subprogrammes (SPs) corresponding to generic research areas to '*Applications*' SPs related to known emerging energy challenges and domains (see Fig. 5). The '*Tools*' SPs include generic research activities involved in most energy technologies around 'new materials', 'physical modelling' and 'characterization'. Presently, there are three identified '*Applications*' SPs, namely 'artificial photosynthesis', 'materials in extreme operating conditions' and 'low temperature heat recovery'. In the future, as the fields addressed in the 'Applications' SPs develop and increase in TRL, they could ultimately lead to dedicated EERA JPs. Alternatively, other 'applications' SPs could be added as they are identified and defined by the

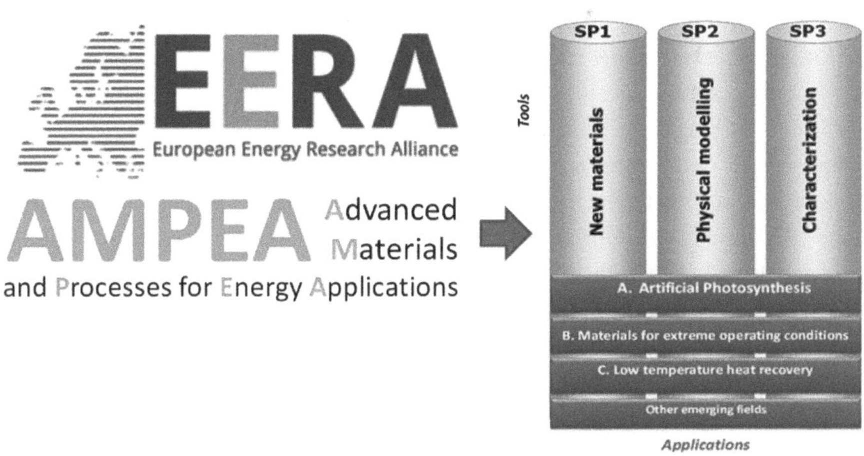

Figure 5 EERA and AMPEA: AMPEA (Advanced Materials and Processes for Energy Applications) is one of the 17 Joint Programmes of the European Energy Research Alliance (EERA). It is focussed on basic science in connection with materials for (emerging) energy applications. AMPEA is structured in subprogrammes organized as a matrix of '*Tools*' (the pillar SPs) and '*Applications*' (transversal SPs). See the text for further details.

AMPEA and more generally by the EERA members. Although focussing on basic science, AMPEA has also developed connections with the industry through establishing a mutual partnership with EMIRI, the Energy Materials Industrial Research Initiative.

The SP on artificial photosynthesis was the first of the transversal '*Applications*' SP launched in AMPEA. It is presently one of the largest in terms of partnership with 18 members, that is, more than half of the total members of the JP. The research programme and objectives developed in the frame of the SP are detailed in the Description of Work (DoW) of this SP. In brief, the aim of this SP 'artificial photosynthesis' is at first to decipher fundamental mechanisms underlying water splitting with the objective to develop devices producing hydrogen with solar-to-hydrogen (STH) efficiencies above 10% on the m^2 scale at low cost. Three types of AP systems are considered in the DoW: Molecularly designed systems, solid state components and nature-guided design systems. All involve a strong development of materials and dedicated advanced characterization tools and techniques supported by theory and modelling, therefore implying a strong interfacing with the related '*Tools*' subprogrammes of AMPEA.

All this was discussed during two workshops organized in the frame of AMPEA. The first one was held in October 2012 at the Max Planck Institute for Chemical Energy Conversion in Mülheim an der Ruhr. It led down the basis of the AMPEA subprogramme on artificial photosynthesis. Shortly afterwards, during the second Joint Programme steering committee held in Berlin in November 2012 (about one year after AMPEA was launched), 15 new organizations mostly interested in Artificial Photosynthesis, joined AMPEA, which turned this SP to be the largest in AMPEA. The main outcomes of this workshop led to the publication of an article by A. Thapper et al. (2013) in the *Green* journal. It was also decided to hold a second dedicated AMPEA cross-linking workshop on artificial photosynthesis. This took place in June 2013, in Wageningen, the Netherlands. Approximately 40 experts from and outside Europe discussed the challenges and presented the views of the AMPEA community in the field of Artificial Photosynthesis in connection with materials, modelling and advanced characterization issues and in line with the general philosophy of the AMPEA JP. This was stated in an updated version of the DoW in early 2014 as challenges and milestones. A detailed discussion on the views of the AMPEA Joint Programme on artificial photosynthesis can be found in an article published by A. Thapper et al. (2013) in the *Green* journal.

Besides workshops, AMPEA promoted the creation of first calls for proposal in artificial photosynthesis and solar fuels within the work programmes of H2020. As a result, these topics are now targeted in some calls of the work programmes for 2016–2017. Some joint proposals involving AMPEA partners were implemented in answer to European calls.

For the future development of AMPEA, the main objectives would be in the future (1) to foster such proposals; (2) to promote the creation of new calls in the forthcoming work programmes of H2020 and beyond in order to support the development of the field. Joint bilateral collaborations can also be promoted by joint calls from two countries. A training mobility scheme with exchange of students and young scientists between partner laboratories could also be a well-suited instrument to foster collaboration between AMPEA partners. Besides, cross-linking AMPEA with other EERA JPs for joint collaborations will also be promoted. In the field of artificial photosynthesis, possible interactions could be developed with the 'bioenergy', 'energy storage' and 'fuel cells and hydrogen technologies' EERA JPs.

To conclude, the 'Integrated Strategic Energy Technology (SET) Plan: Accelerating the European Energy System Transformation' released in September 2009 by the European Commission with a strengthened role of the EERA and its Joint Programmes will be an excellent framework for fostering the development of artificial photosynthesis and solar fuels.

3.2 Other International Initiatives

3.2.1 International Energy Agency

Since 1977, the International Energy Agency http://www.iea.org/ (IAE), an autonomous body within the Organization for Economic Cooperation and Development (OECD), has set a Hydrogen Implementing Agreement (HIA, 1977), which pursues collaborative hydrogen research and development and information exchange among its member countries, in order to accelerate hydrogen implementation and its widespread utilization for optimal environmental protection, energy security and international economic development. The HIA gathers partners from 20 countries (Australia, Belgium, Denmark, Finland, France, Germany, Greece, Japan, Israel, Italy, Korea, Lithuania, the Netherlands, New Zealand, Norway, Spain, Sweden, Switzerland, United Kingdom and United States) plus the Commission of the European Union and the United Nations Industrial Development Organization (UNIDO).

More precisely, HIA aims at coordinating hydrogen research, development and demonstration activities on a global basis among member countries through the creation and conduct of tasks or annexes. Remarkably, eight of the 40 tasks implemented within the HIA over the 38 years of operation concerned light-driven hydrogen production (HIA, 1977). Since 1999, the topic is divided up into two tasks running in parallel, one devoted to biological and bioinspired molecular systems (bioinspired artificial photosynthesis and biological hydrogen production, tasks 15, 21) and another one to physicochemical processes related to photoelectrochemical water splitting (tasks 14, 20 and 26). A. Luzzi (Institut für Solartechnik, Rapperswil, Switzerland) and E. L. Miller (US Department of Energy) acted as operating agents to coordinate task 14 (photoelectrolytic production of hydrogen, 1999–2004), task 20 (hydrogen from waterphotolysis, 2004–2007) and task 26 (advanced materials for waterphotolysis, 2008–2013). Specific materials topics covered within task 26 included (Miller, 2012): (1) III–V crystalline semiconductors material systems, (2) Fe_2O_3-based thin-film materials, (3) WO_3-based thin-film materials, (4) I–III–VI_2 thin-film semiconductor materials (eg, $CuGaSe_2$), (5) molybdenum disulfide (MoS_2) nanostructured photocatalysts, (6) bismuth vanadate ($BiVO_4$) materials, (7) tantalum oxinitride (TaON) materials, (8) PEC materials theory updates and (9) new oxide materials and material systems.

The 2015–2020 HIA strategic plan now includes photoelectrochemical processes together with renewable electrolysis and solar-thermochemical water-splitting in task 35. The work devoted to bioinspired and biological hydrogen production is now pursued within task 34 (Biological H_2 for Energy and Environment).

3.2.2 Solar Fuels Institute

The Solar Fuels Institute (SOFI) was initiated in 2011 (SOFI, 2011). It is an international global research consortium of universities, government laboratories and industries united around the goal of developing and commercializing a liquid solar fuel. Presently, SOFI gathers 19 members from 12 countries all over the world. These include universities (eg, Northwestern University, Uppsala University, TU Delft) and research centres (eg, JCAP, Argonne National Laboratory, DIFFER, MPI for Chemical Energy Conversion), RTOs (CEA), national centres and initiatives on artificial photosynthesis (eg, Swedish Consortium for Artificial Photosynthesis, Korea Centre for Artificial Photosynthesis, UK Solar Fuels Network) as well as industries (eg, Total, Shell, IL&FS). SOFI leverages substantial public and private commitment to energy sustainability in order to create global industry—academic partnerships to achieve the long-term vision and to explore opportunities to commercialize interim technologies along the way. SOFI adopts a 'team science' approach to overcome the technical, institutional and behavioural hurdles to innovation. Using advanced collaborative tools, SOFI reaches beyond the capacity of any individual research laboratory or institution. Thus SOFI solves the problem of mass coordination by aggregating, organizing, informing, focussing and – most importantly – empowering the world's innovators to work together as an integrated unit to tackle and solve the problem of scaled, commercial production of liquid fuels from sunshine.

SOFI's primary missions can therefore be summarized as follows:

- Identify those technologies most likely to find commercial purchase by inviting industry partners to explore SOFI research and establish milestones on the road to translational development.
- Bridge the 'valley of death' between basic research funding, largely the domain of governments and philanthropy and later stage technology investment, attractive to industry and venture capital by exploring innovative ways of financing translational research such as prizes, public/private partnerships and alternative funding models.
- Overcome an entrenched academic culture that measures success by the number of papers published in prestigious journals, by partnering with funding agencies in changing performance metrics.
- Create a collaborative platform that allows for and encourages the pooling of resources and sharing of knowledge among research labs and industries.

In order to fulfil these missions, SOFI has launched several initiatives and tools such as the SOFI Fellows Exchange Program and the SOFI Knowledge Map.

3.2.2.1 SOFI Fellows Exchange Program

SOFI helps researchers (graduate students, postdoctoral associates, staff scientists and engineers and principal investigators) to carry out collaborative projects in another institution's lab(s), member of SOFI. These guest researchers are designated 'Solar Fuels Institute Fellows.' Through exchanges that range from 2- to 12-week duration, SOFI breaks down the silos in solar fuels research and builds a multidisciplinary network, while addressing specific project needs of the laboratories. The SOFI Fellows Program lowers many of the barriers to carrying out collaborative projects across geographically distributed teams. SOFI has identified six cross-cutting developmental *foci* (fully integrated device, photoanode, photocathode, electrolyser, benchmarking and techno-economic analysis) that comprise overlapping technologies. SOFI seeks to facilitate Lab Exchange projects that integrate various system components from different laboratories, thus educating a workforce adapted to solve problems at the intersections of current knowledge. Examples of such exchanges supported by SOFI to date include bilateral projects between Arizona State University and CEA-Grenoble; between Uppsala University and University of California, San Diego; and between Imperial College London and Northwestern University.

3.2.2.2 SOFI Knowledge Map

Comparing research results is often an apples to oranges situation. Science relevant to the development of solar fuels is spread across multiple disciplines and industries. The lack of common metrics and benchmarking hinders innovation. The SOFI Knowledge Map is a web-based platform that uses unique solar fuels ontology to facilitate search and discovery. The Knowledge Mapping process identifies and categorizes knowledge assets within SOFI and the solar fuels research areas – people, processes, content and technology. This process enables SOFI to leverage existing knowledge resident within the SOFI network, recruit new expertise as needed and identify barriers and constraints to fulfilling strategic goals and objectives. The goal of the Knowledge Map is to connect information and knowledge with the right people at the right time, and to separate useful results and data from publication PDFs such that they are digitally accessible to the community via a program application interface (API).

The publication portal of the SOFI Knowledge Map was launched in April 2015. It contains 61,000 peer-reviewed publications from 100 countries and nearly 3000 journal titles as of August 2015. Covering core areas such as Light Capture, Catalysis and Device Engineering, the SOFI Knowledge Map provides intuitive access to relevant data in a standardized way, speeding innovation and discovery by collecting the fragmented solar fuels literature into a coherent, searchable knowledge database.

For publications in the future, SOFI is partnering with the major publishers so that the data input is part of the manuscript submission process. As such, the SOFI Knowledge Map will be a data repository similar to the Cambridge Crystallographic Data Centre where authors voluntarily submit their results, performance metrics and measurements to an external database.

3.2.3 *Miscellanea*

In the previous sections, we described international initiatives associating partners from different countries outside of Europe. In the following, we give a brief (but nonexhaustive) overview of national initiatives on solar fuels in countries outside of Europe (Faunce et al., 2013).

In United States, the Office of Basic Energy Sciences in the Department of Energy's Office of Science established the EFRC programme to accelerate transformative discovery (EFRC, 2015). In 2009 5-year awards were made to 46 EFRCs, and an open re-competition of the programme in 2014 resulted in 4-year awards to 32 centers, 22 of which were renewals. These integrated, multiinvestigator Centres unite universities, national laboratories, nonprofit organizations and companies to conduct fundamental research focussing on one or more 'grand challenges'. Solar fuels research is a major focus for two to three EFRCs, and research activities are present in many others. Below are a few highlights:

- The Center for Solar Fuels based at the University of North Carolina at Chapel Hill focuses on the concept of the dye-sensitized photoelectrosynthesis cell (DSPEC) for solar fuels production. Research on high efficiency DSPEC devices in the Centre for Solar Fuels builds on a modular approach: catalysts for water oxidation or CO_2 reduction, light-harvesting chromophores and chromophore arrays, chromophore-catalyst assemblies, mesoporous nanoparticle semiconductor oxide and transparent conducting oxide films and core/shell structures are all investigated separately and integrated in thin-film structures for possible DSPEC applications.
- The Argonne-Northwestern Solar Energy Research (ANSER) Centre emphasizes catalysts ranging from molecules, to clusters, to nanoparticles, to mesoscale materials that have promising compositions that are thermodynamically inaccessible as bulk (macroscopic) materials. The solar fuels research at the ANSER Centre is focussed on (1) organic/inorganic/hybrid light harvesters, (2) water oxidation catalysis, (3) water reduction catalysis and (4) carbon dioxide reduction catalysis.
- The Centre for Biological Electron Transfer and Catalysis (BETCy) led by Montana State University investigates the mechanisms and structural basis controlling electron bifurcation, electron–ion coupling, and redox catalysis in model enzymes. The detailed understanding of how biological systems control and exploit electron flow is used to develop modular biochemical conversions for the production of hydrocarbon and hydrogen biofuels.

- The Centre for Molecular Electrocatalysis (CME) based at the Pacific Northwest National Laboratory is designing and creating molecular electrocatalysts for the reduction of protons (hydrogen generation), reduction of O_2 and the reduction of nitrogen to ammonia. One focus is to understand and control proton relays and transfers, as they play critical roles in fuel-forming chemical transformations that are multiproton and multielectron reactions.

The DOE also created Energy Innovation Hubs in 2010. Among them, the JCAP has been established in 2010 for a period of 5 years and received a funding of $122 million. It is led by the California Institute of Technology (Caltech) together with its main partner, the Lawrence Berkeley National Laboratory (LBNL) and gathers 150 scientists and engineers (JCAP, 2010). It also has partnership with other US academic partners in California: the University of California campuses at Irvine (UCI) and San Diego (UCSD), as well as the Stanford Linear Accelerator (SLAC) and generally acts as a central hub for other solar-fuels research teams in United States. Currently, it is the world's largest research programme in artificial photosynthesis. In April 2015, the Department of Energy (DOE) announced a 5-year, $75 million renewal of JCAP.

In Asia, the Japanese government launched a World Premier International Research Centre Initiative for establishing in Japan world-class Centres active on basic multidisciplinary research domains based on calls in 2007, 2010 and 2012 (WPI, 2007). Several of the successful Centres address artificial photosynthesis, such as the International Institute for Carbon-Neutral Energy Research at Kyushu University, the Advanced Institute for Materials Research at Tohoku University or International Centre for Materials Nanoarchitectronics. In 2012, the national project, 'All Nippon Artificial photosynthesis Project for Living Earth (AnApple)' was launched as one of the Scientific Researches on Innovative Areas funded for 5 years. More than 40 Japanese leading scientific groups joined this project.

Korea and China have also prominent research programmes and research centres focussing on artificial photosynthesis. South Korea hosts at Sogang University since 2009 the Korean Centre for Artificial Photosynthesis KCAP (2009). KCAP has established a Memorandum of Understanding with the Helios-Solar Energy Research Centre (SERC) in Lawrence Berkeley National Laboratory (LBNL) in United States. In China, the National Laboratory for Clean Energy in Dalian (DNL) was recently inaugurated and has, within the 'Clean energy Division' a research resource centre on 'Hydrogen energy and advanced materials' where artificial photosynthesis and solar fuels are addressed (DNL).

In addition, all around the world, other ambitious programme initiatives on artificial photosynthesis and solar fuels have been launched either in the abovementioned countries or in other ones around the world, for example in Singapore at the Nanyang Technological University with the Solar Fuels Lab (NTU Solarfuels) or in Australia with the Australian Centre of Excellence on Electromaterials Science (ACES) established by the Australian Research Council and led by the University of Wollongong (ACES, 2014).

Yet, this list does not pretend to be exhaustive and it can be foresighted that with the increasing interest for solar fuels in the context of the transition to a low carbon energy society, new ambitious national and international programmes and initiatives will be launched in the future.

4. CONCLUSION AND PERSPECTIVES

This overview of the field of 'artificial photosynthesis' and 'solar fuels' reveals that the collaborative schemes are so far mostly developed at the national level. This probably reflects funding scenarios but we note that even the various European Research Frameworks with collaborative calls opened since 1994 have not supported as required the building of an integrated European community in the field. Bilateral calls have nevertheless been launched, for example between China and Sweden, or France and Japan, and such initiatives should be pursued and encouraged. Indeed, with funding missing at the international level, the extent of collaborative networks, which have been created though, remains limited to the organization of meetings where concepts and results can be discussed. The possibility to create an international society for Solar Fuels has been discussed in the frame of meetings of AMPEA, SOFI or within ISF conference organization committee. This point is so far postponed and will depend on the success of the second ISF conference. In a more global context, the question of a Global Project on Artificial Photosynthesis has been discussed in the context of meetings organized by T. Faunce. 'Artificial Photosynthesis' is here proposed as a new worldwide adopted policy globalizing practical technologies to address climate change and energy and food security concerns (Faunce, 2012, 2015; Faunce et al., 2013).

ACKNOWLEDGEMENTS

The authors would like to thank Hervé Bercégol, Sebastian Fiechter and Stenbjörn Styring for stimulating discussions and support when preparing this chapter. The support of Magali Le Discorde (BioDoc Service, Life Science Division, CEA Saclay) and Marie-Anne Leriche

(Mission Bibliométrie, Matter Science Division, CEA Saclay) during the bibliometric analysis is gratefully acknowledged. This work has been supported by the COST Action CM1202, the French National Research Agency (Labex program, ARCANE, ANR-11-LABX-0003-01 and the European Energy Research Alliance (EERA) joint program AMPEA. SOFI acknowledges support from its members and is grateful for the guidance and insights from the many dozens of researchers around the globe who have generously offered their time and wisdom in pursuit of the vision of bringing artificial photosynthesis to market for the common good.

REFERENCES

ACES. (2014). *Australian Centre of Excellence on Electromaterials Science*. http://www.electromaterials.edu.au/.

Ager, J. W., Shaner, M. R., Walczak, K. A., Sharp, I. D., & Ardo, S. (2015). Experimental demonstrations of spontaneous, solar-driven photoelectrochemical water splitting. *Energy and Environmental Science, 8*(10), 2811–2824.

AMPEA. (2011). *Advanced Materials and Processes for Energy Applications (AMPEA) EERA Joint Programme*. www.eera-set.eu/eera-joint-programmes-jps/advanced-materials-and-processes-for-energy-application-ampea/.

AnApple. (2012). *All Nippon Artificial Photosynthesis Project for Living Earth (AnApple)*. www.artificial-photosynthesis.net/.

BioSolarCells (2009). www.biosolarcells.nl/en/home.html.

Blankenship, R. E., Tiede, D. M., Barber, J., Brudvig, G. W., Fleming, G., Ghirardi, M., … Sayre, R. T. (2011). Comparing photosynthetic and photovoltaic efficiency and recognizing the potential for improvement. *Science, 332*, 805–809.

Bolton, J. R. (1978). Solar fuels. *Science, 202*, 705–711.

Campagna, S., Bonchio, M., Venturi, M., & Scandola, F. (August 2012). Verso la fotosintesi artificiale: competenze, strutture e progetti di ricerca nel settore. *La Chimica and l'Industria*, 88–93.

DNL: Dalian National Laboratory for Clean Energy. www.dnl.dicp.ac.cn/index_en.php.

EERA. (2008). *European Energy Research Alliance (EERA)*. www.eera-set.eu.

EFRC. (2015). *Energy Frontier Research Centres*. http://science.energy.gov/bes/efrc/.

EMIRI: Energy Materials Industrial Research Initiative. https://emiri.eu/.

Communication from the Commission to the European Parliament, the Council, the European Economic and Social Committee and the Committee of the Regions – Energy Roadmap 2050/ COM/2011/0885 final */*.(2013). Available from: http://eur-lex.europa.eu/legal-content/EN/NOT/?uri=celex:52011DC0885.

Eurostat shares 2013, short assessment of renewable energy sources (2013). Available from: http://ec.europa.eu/eurostat/web/energy/data/shares.

Faunce, T. A., Lubitz, W., Rutherford, A. W., MacFarlane, D., Moore, G. F., Yang, P., … Styring, S. (2013). Energy and environment policy case for a global project on artificial photosynthesis. *Energy and Environmental Science, 6*, 695–698.

Faunce, T. (2012). Towards a global solar fuels project- Artificial photosynthesis and the transition from anthropocene to sustainocene. *Procedia Engineering, 49*, 348–356.

Faunce, T. (2015). Does the world need a global project on artificial photosynthesis? *Interface Focus, 5*, 20150029. http://dx.doi.org/10.1098/rsfs.2015.0029.

GDR Solar Fuels, (2016). www.solar-fuels.cnrs.fr.

GRC Solar fuels. (2007). *Gordon Research Conference (GRC) and Gordon Research Seminar (GRS) series "Renewable Energy: Solar Fuels"*. https://www.grc.org/conferences.aspx?id=0000556.

Helmholtz Society Institute of Solar Fuels. http://www.helmholtz-berlin.de/forschung/oe/ee/solare-brennstoffe/index_en.html.

Hydrogen Implementing Agreement (HIA) (1977) . http://ieahia.org/. The list of 40 tasks implemented in the frame of the HIA can be consulted at http://ieahia.org/page.php?s=glance&p=tasks.

INTELLIXIR©. http://intellixir.com/.

ISF 1. (2015). *International Solar Fuels Conference ISF-1, April 26-May 1, 2015, Uppsala, Sweden*. www.solarfuel.se/meetings/.

JCAP. (2010). *Joint Centre for Artificial Photosynthesis*. www.solarfuelshub.org.

KCAP. (2009). *Korean Centre for Artificial Photosynthesis*. http://www.k-cap.or.kr/eng/info/index.html?sidx=1.

Lewis, N. S., & Nocera, D. G. (2006). Powering the planet: chemical challenges in solar energy utilization. *Proceedings of the National Academy of Sciences of the United States of America, 103*, 15729–15735.

Miller, E. L. (2012). IEA-HIA Task 26 research and development progress in renewable hydrogen production through photoelectrochemical water splitting. *Energy Procedia, 29*, 438–444.

MPI CEC. (2012). *Max Planck Institute for Chemical Energy Conversion*. http://www.cec.mpg.de/home.html?L=1.

Solar Fuels Lab, Nanyang Technological University in Singapore: http://www.mse.ntu.edu.sg/FacilitiesEquipment/ResearchFocusLab/Pages/ArtificialPhotosynthesis.aspx.

PECDEMO. (2015). *Photoelectrochemical Demonstrator Device (PECDEMO) European project*. http://pecdemo.epfl.ch/.

PECHouse 2, (2015) Photoelectrochemistry Competence Centre PECHouse 2: http://pechouse.epfl.ch/

Perspect-H_2O. (2012). *COST action Perspect-H_2O*. http://www.perspect-h2o.eu/.

SET Plan. (2008). *Strategic Energy Technology Plan (SET Plan)*. https://ec.europa.eu/energy/en/topics/technology-and-innovation/strategic-energy-technology-plan.

SFN. (2012). *Solar Fuels Network*. http://solarfuelsnetwork.com/events-archive/.

SOFI Knowledge Map. https://km.solar-fuels.org/.

SOFI. (2011). *Solar Fuels Institute (SOFI)*. http://www.solar-fuels.org/.

SolarCap. (2009). *SolarCAP Consortium for Artificial Photosynthesis*. www.solarcap.org.uk.

Swedish Consortium. (1994). *Swedish Consortium for Artificial Photosynthesis*. http://www.solarfuel.se/.

Thapper, A., Styring, S., Saracco, G., Rutherford, A. W., Robert, B., Magnuson, A., … Antero, V. (2013). Artificial photosynthesis for solar fuels – an evolving research field with AMPEA, a joint programme of the European Energy Research Alliance. *Green, 3*, 43–57.

Thompson Reuters Web of Science database. Can be accessed from: http://apps.webofknowledge.com/.

Wood, C. J., Summers, G. H., Clark, C. A., Kaeffer, N., Braeutigam, M., Carbone, L. R., … Gibson, E. A. (2016). A comprehensive comparison of dye-sensitized NiO photocathodes for solar energy conversion. *Physical Chemistry Chemical Physics*. http://dx.doi.org/10.1039/c5cp05326a.

WPI. (2007). *World Premier International Research Centre Initiative*. http://www.jsps.go.jp/english/e-toplevel/11_gaiyo.html.

SUBJECT INDEX

'*Note:* Page numbers followed by "f" indicate figures and "t" indicate tables.'

A

B

F

H

I

Q

R

AUTHOR INDEX

Note: Page numbers followed by "f" indicate figures.

C

E

F

G

H

M

N

Q

R

S

T

U

V

W

X

Y

Z

FSC
www.fsc.org
MIX
Papier | Fördert
gute Waldnutzung
FSC® C083411

Kolibri 360 GmbH
Ferdinand-Jühlke-Straße 7
99095 Erfurt, Deutschland
produktsicherheit@kolibri360.de